Introducing Tectonics, Rock Structures and Mountain Belts

Companion Titles

Introducing Geology – A Guide to the World of Rocks (Second Edition 2010)
Introducing Palaeontology – A Guide to Ancient Life (2010)
Introducing Volcanology – A Guide to Hot Rocks (2011)
Introducing Geomorphology – A Guide to Landforms and Processes (2012)
Introducing Meteorology (forthcoming 2012)
Introducing Oceanography (forthcoming 2012)

For further details of these and other Dunedin
Earth and Environmental Sciences titles see
www.dunedinacademicpress.co.uk

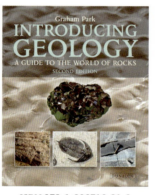

ISBN 978-1-906716-21-9

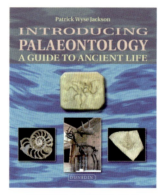

ISBN 978-1-906716-15-8

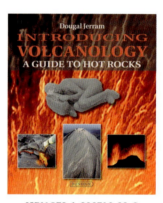

ISBN 978-1-906716-22-6

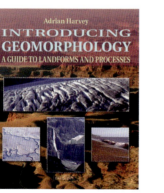

ISBN 978-1-906716-32-5

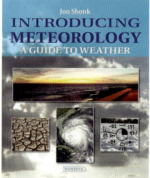

ISBN 978-1-780460-02-4

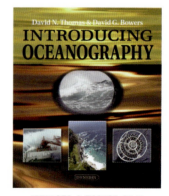

ISBN 978-1-780460-01-7

Introducing Tectonics, Rock Structures and Mountain Belts

Graham Park

Published by
Dunedin Academic Press Ltd
Hudson House
8 Albany Street
Edinburgh EH1 3QB
Scotland

www.dunedinacademicpress.co.uk

ISBN 978-1-906716-26-4

British Library Cataloguing in Publication data
A catalogue record for this book is available from the British Library

Design and pre-press production by Makar Publishing Production, Edinburgh
Printed in Poland by Hussar Books

Contents

Acknowledgements

I am indebted to Professor John Winchester and an anonymous reviewer for many helpful comments and suggestions that have resulted in significant improvements to this book.

I also wish to thank my wife Sylvia for her unfailing support and for subjecting the manuscript to the valuable scrutiny of a non-geologist.

List of illustrations and tables

Tables

Sourced illustrations

The following illustrations are reproduced by permission.
British Geological Survey. ©NERC. All rights reserved. IPR/73-34C: Figure 2.7
Jim Wark airphotos: Figures 5.6, 6.1, 6.8
NASA photo courtesy Space Images: Figure 2.1
NASA photo courtesy Parstimes: Figures 9.5A, B.

The following illustrations have been adapted from published sources.
Figure 2.2: Wyllie, P.J. 1976. *The way the Earth works*, Wiley, New York.
Figure 2.3: Hamblin, W.K. 1989. *The Earth's dynamic systems*, 5th edition. Macmillan, New York.
Figure 2.5: Wyllie, P.J. 1976. *The way the Earth works*, Wiley, New York.
Figure 2.8: Zeuner, F.E. 1958. *Dating the past* (4th edn) Methuen, London.
Figure 3.2: Hamblin, W.K. 1989. *The Earth's dynamic systems*, 5th edition. Macmillan, New York.
Figure 3.3: McElhinny, N.W. 1973. *Palaeomagnetism and plate tectonics*, Cambridge University Press.
Figure 3.6: Larson, R.L. & Pitman, W.C. 1972. *Bulletin of the Geological Society of America*, 83, 3645–3661.
Figure 3.7: Vine, F.J. & Hess, H.H. 1970, *The Sea*, Wiley, New York.
Figure 3.9A: Bott, M.P. *The interior of the Earth*, Arnold, London.
Figure 3.9B: Girdler, R.W. & Darracott, B.W. 1972. African poles of rotation. *Comments on the Earth Sciences: Geophysics* 2, (5), 7–15.
Figure 3.11: Uyeda, S. 1978. *The new view of the Earth*. Freeman, San Francisco.
Figure 3.12: Morgan, W.J. Deep mantle convection plumes and plate motions. *Bulletin of the American Association of Petroleum Geologists*, 56, 203–213.
Figure 3.13: Saemundsson, K. 1974. Evolution of the axial rifting zone in northern Iceland. *Bulletin of the Geological Society of America*, 85, 495–504.
Figure 5.4D: Reston, T.J. 2007. The formation of non-volcanic rifted margins by the progressive extension of the lithosphere: the example of the West Iberian margin. From: Karner, G.D., Manatscheal, G. & Pinheiro, L.M. (eds) *Imaging, mapping and modeling continental lithosphere extension and breakup*. Geological Society, London, Special Publications, 282, 77–110.
Figure 5.5A: Virtual Seismic Atlas, Robert Butler.
Figure 5.6: Reston, T.J. 2007 (see above).
Figure 5.7C: Elliott, D & Johnston, M.R.W. 1980. Structural evolution in the northern part of the Moine thrust zone. *Transactions of the Royal Society of Edinburgh: Earth Sciences*, 71, 69–96.
Figure 5.9B: Park, R.G. 1961. The pseudotachylite of the Gairloch district, Ross-shire, Scotland. *American Journal of Science*, 259, 542–550.
Figure 6.13: Ramsay, J.G. 1980. Shear zone geometry: a review. *Journal of Structural Geology*, 2, 83–99.
Figure 8.2C: McGregor, M.D. & McGregor, A.G. 1948. *British Regional Geology: the Midland Valley of Scotland*. HMSO, London.
Figure 8.2E: Nicholson, R. & Pollard, D.D. 1985. Dilation and linkage of en-echelon cracks. *Journal of Structural Geology* 7, 583–590.
Figure 8.4: Johnstone, G.S. 1966. *British Regional Geology: the Grampian Highlands (3rd Edn)*. HMSO, London.
Figure 8.5: Hutton, D.H.W. 1988. Granite emplacement mechanisms and tectonic controls: inferences from deformation studies. *Transactions of the Royal Society of Edinburgh: Earth Sciences*, 79, 245–255.
Figure 8.6: Emeleus, C.H. & Bell, B.R. 2003. *Scotland: the Tertiary volcanic districts. British Regional Geology, 4th Edition*, London, HMSO for the British Geological Survey.
Figure 9.2: Graham, R.H. 1981. Gravity sliding in the Maritime Alps. From: McClay, K.R. & Price, N.J. (eds) *Thrust and nappe tectonics*. Geological Society, London, Special Publication 9, 335–352.
Figure 9.4: Trusheim, F. 1960. Mechanism of salt migration in Germany. *Bulletin of the American Association of Petroleum Geologists*, 44, 1519–1540.
Figure 10.1: Hatcher, R. & Williams, R.T. 1986. Mechanical model for single thrust sheets. *Geological Society of America Bulletin*, 97. 975–985.
Figure 10.2: Westbrook, G.K. 1982. The Barbados Ridge complex: tectonics of a mature forearc system. In: Leggett, J.K. (ed.) *Trench-forearc geology: sedimentation and tectonics on modern and ancient active plate margins*. Geological Society, London, Special Publications 10, 275–290.
Figure 10.4. Godin, L., Grujic, D., Law, R.D. & Searle, M.P. 2006. Channel flow, ductile extrusion and exhumation in continental collision zones: an introduction. From: Law, R.D., Searle, M.P. & Godin, L. (eds) *Channel flow, ductile extrusion and exhumation in continental collision zones*. Geological Society, London, Special Publication 268, 1–23.
Figure 11.1: Smith, A.G. & Briden, J.C. 1977. *Mesozoic and Cenozoic palaeocontinental maps*. Cambridge University Press.
Figure 11.2A. Molnar, P. & Taponnier, P. 1975. Cenozoic tectonics of Asia: effects of a continental collision. *Science* 189, 419–426.
Figure 11.3A, B. Harrison, T.M. 2006. Did the Himalaya crystallines extrude partially molten from beneath the Tibetan plateau? From: Law, R.D., Searle, M.P. & Godin, L. (eds) *Channel flow, ductile extrusion and exhumation in continental collision zones*. Geological Society, London, Special Publication 268, 237–254.

Figure 11.3D, 11.4: Searle, M.P, Elliott, J.R, Phillips, R.J. et al. 2011. Crustal-lithospheric structure and continental extrusion of Tibet. *Journal of the Geological Society, London*, 168, 633–672.

Figure 11.5A. Coward, M.P. & Dietrich, D. 1989. Alpine tectonics: an overview. From: Coward, M.P., Dietrich, D. & Park, R.G. (eds) *Alpine tectonics*. Geological Society, London, Special Publication 45, 1–29.

Figure 11.7. Ramsay, J.G. 1963. Stratigraphy, structure and metamorphism in the Western Alps. *Proceedings of the Geologists Association* 74, 357–392.

Figure 11.8A. Coney, P.J., Jones, D.L. & Monger, J.W.H. 1980. Cordilleran suspect terranes. *Nature, London*, 288, 329–333.

Figure 12.1: Dalziel, I.W.D. 1997. Neoproterozoic-Palaeozoic geography and tectonics: review, hypothesis, environmental speculation. *Geological Society of America Bulletin*, 109, 16–42.

Figure 12.3A: Elliott, D. & Johnson, M.R.W. 1980. Structural evolution in the northern part of the Moine thrust zone. *Transactions of the Royal Society of Edinburgh: Earth Sciences*, 71, 69–96.

Figure 12. 3B: Treagus, J.E. 2000. *Solid geology of the Schiehallion district*. Memoir of the British Geological Survey, HMSO.

Figure 12.3D: Leggett, J.K., McKerrow, W.S. & Eales, M.H. 1979. The Southern Uplands of Scotland: a Lower Palaeozoic accretionary prism. *Journal of the Geological Society, London*, 136, 755–770.

Figure 12.3F: Coward, M.P. & Siddans, A.W.B. 1979. The tectonic evolution of the Welsh Caledonides. From: Harris, A.L. Holland, C.H. & Leake, B.E. (eds) *The Caledonides of the British Isles – reviewed*. Geological Society of London, Special Publication 8, 187–198.

Figure 12.5: Dewey, J.F. & Shackleton, R.M. 1984. A model for the evolution of the Grampian tract in the early Caledonides and Appalachians. *Nature, London*, 312, 115–121.

Figure 12.6. Pisarevsky, S.A., Wingate, M.T.D., Powell, C. McA., Johnson, S., Evans, D.A.D., 2003. Models of Rodinia assembly and fragmentation. From: Yoshida, M., Windley, B., Dasgupta, S. (eds) *Proterozoic East Gondwana: supercontinent assembly and breakup*. Geological Society, London, Special Publication 206, 35–55.

Figure 12.7, 12.8: Rivers, T. The Grenville Province as a large, hot, long-duration collisional orogen: insights from the spatial and thermal evolution of its orogenic fronts. From: Murphy, J.B., Keppie, J.D. & Hynes, A.J. 2009. *Ancient orogens and modern analogues*. Geological Society of London, Special Publication 327, 405–444.

Figure 12.9, 12.10: Corrigan, D., Pehrsson, S., Wodicka, N. & de Kemp, E. 2009. The Palaeoproterozoic Trans-Hudson Orogen: a prototype of modern accretionary processes. From: Murphy, J.B., Keppie, J.D. & Hynes, A.J. 2009. *Ancient orogens and modern analogues*. Geological Society of London, Special Publication 327, 457–479.

Figure 12.11: Card, K.D. & Cieselski, A. 1986. DNAG#1: Subdivision of the Superior Province of the Canadian Shield. *Geoscience Canada* 13, 5–13.

Figure 12.12: Friend, C.R.L. & Nutman, A.P. 2005. New pieces to the Archaean terrane jigsaw puzzle in the Nuuk region, southern West Greenland: steps in transforming a simple insight into a complex regional tectonothermal model. *Journal of the Geological Society, London*, 162, 147–162.

Preface

This book is not intended to be a textbook, but is designed to explain the key concepts of tectonics and rock structures to both students and others interested in geology – especially those who may not have a good scientific or mathematical background. The study and understanding of geological structures has traditionally been guided by the rigorous application of mathematics and physics, and conventional textbooks on structural geology have reflected this approach. However, in my experience, many students are discouraged by this aspect of the subject and consequently, in this book, I have avoided mathematical equations altogether, and reduced the geometry to the minimum I judged necessary to understand the concepts. Those who wish to gain a deeper understanding of the subject, or who are engaged on a university-level course in structural geology, are directed to the reading list, which contains several excellent textbooks and on-line sources recommended for further study.

The application of plate tectonic theory has revolutionised structural geology by giving the study of structures a context in which they can be explained. Since the large-scale movements of the plates ultimately control smaller-scale structures, the study of tectonics is therefore the key to understanding the latter. I therefore introduce the reader first to large-scale Earth structure and the theory of plate tectonics. The following four chapters deal in rather more detail with what might be called 'traditional' structural geology – the study of the response of rock material to crustal forces, and the explanation for the bewildering variety of rock structures formed thereby. This aspect of geology was transformed in the 1950s and 1960s by rigorous geometrical analysis and the application of the techniques of rock mechanics. I believe that it is important for the student of geology to be aware of this background, without necessarily being able (at least initially!) to understand the mathematical or geometrical detail.

An important development in the latter decades of the last century was the emphasis by structural geologists on the movement history of rock masses relative to each other as revealed by the study of fault systems and shear zones. This enabled structural geology to be more directly related to plate tectonics and helped to integrate geological structures with tectonics; I have tried to emphasise this aspect in the book.

One of the most exciting aspects of geology is the study of the great mountain ranges ('orogenic belts'), both of the present day and of the past. The final three chapters of the book attempt to explain how knowledge of plate-tectonic theory, geological structures and the processes of deformation may be employed to understand these orogenic belts.

Scientists are addicted to the classification and naming of the things that fall into their domain, and structural geologists are no exception. However, I have tried to avoid the excessive use of terminology, and all technical terms are defined and explained in the Glossary, which readers are encouraged to check when uncertain of the exact meaning of a word used in a geological context (which often differs from its everyday meaning!).

The Appendix contains a set of four tables that set out in a simplified way the main terminology of the geological timescale and of igneous, sedimentary and metamorphic rocks as an aid to those readers who may not be so familiar with these aspects of the subject.

Finally, I hope that the reader might share the excitement of discovering how the all-embracing theory of plate tectonics can help to explain the multitude of complexities revealed in the study of the rocks.

R.G.P.
December 2011

Note: all terms highlighted in **bold** are defined in the Glossary at the end of the book.

1   Introduction

Meaning and scope of the terms 'tectonic' and 'structure'

The adjective 'tectonic' merely means 'structural', i.e. *applying to a structure*. However, in geological usage it has come to be applied particularly to large-scale structures – thus 'tectonic plate'. In order to explain rock structures, it is necessary to understand the forces operating within and on the crust that are responsible for creating them, and to do this, we need to investigate the Earth-scale processes known collectively as plate tectonics.

The term 'structure' in everyday usage refers normally to a building or other artificial construction, but in geological terms it has come to mean a body of rock whose shape can be defined geometrically and which has originated by a geological process. The most obvious and best known types of geological structure are folds and faults, which have been produced by the action of geological forces within the Earth's crust and which can give us an insight into the magnitude of these forces and how they operate. Another group of structures is typical of deep-seated metamorphic rocks where crustal forces have effected thoroughgoing changes in the rock, producing new textures through recrystallisation; the structure produced by such changes is known as the fabric. A fourth category of geological structures consists of igneous intrusions. Such bodies are influenced by the forces acting within the crust during their emplacement and the igneous bodies themselves may cause structural changes in the host rocks.

Kinematic and dynamic models of deformation

Kinematics is the study of *movement* and dynamics the study of *forces*, and both types of model have been used in the investigation of geological structures. Dynamic models have traditionally been used by structural geologists employing the techniques of mechanical engineering and materials science. This requires a familiarity with mathematics and physics, which is a potential barrier to understanding for many students. However, whereas a background in these subjects is essential to understanding the deformation of rock in any depth, it is not really necessary to be able to grasp the essentials of structural geology at a basic level.

The kinematic model of deformation has become in many ways more popular as a way of understanding geological structures. In this type of approach, structures are analysed in terms of the relative movement between blocks of crust, and the causal forces are disregarded. Thus, systems of related folds and faults are explained by a single set of relative movements. Such a system can also be integrated into the plate tectonic model by scaling up to a higher order of magnitude but maintaining the same overall movement sense. Examples of this are discussed in Chapters 10–12, where we look in detail at the structure of orogenic belts. At this larger scale, structures are grouped into tectonic regimes, based on their over-riding characteristics, and linked with their plate tectonic setting, thus: extensional regimes are associated with divergent plate boundaries (continental rift zones); compressional regimes with convergent plate boundaries (subduction and collision zones); and strike-slip regimes with transform faults. These plate boundary types are explained in Chapter 3.

Layout of the book

The sequence of chapters reflects the author's belief that an understanding of the plate tectonic model and the evidence for it is helpful in the understanding of geological structures on the smaller scale. Thus Chapters 2–3 deal with large-scale Earth structure and the theory of plate tectonics. These chapters are followed by chapters 4–9 on the process of deformation and on the various types of geological structure. Finally, Chapters 10–12 deal with the study and understanding of orogenesis and orogenic belts. A brief summary of each chapter follows.

◆ *Chapter 2: Large-scale earth structure.* This chapter describes the most significant features of the Earth's crust: the continents and oceans, mountain belts, ridges and trenches; with the Earth's internal structure, the **asthenosphere** and **lithosphere**; and finally with the distribution of present-day tectonic activity in the form of earthquakes and volcanoes.

◆ *Chapter 3: Plate tectonics.* Modern plate tectonics starts with the theory of **continental drift**, which, when added to the more recent concept of **ocean-floor spreading**, led in the 1960s to the development of **plate tectonics**. The concept of the tectonic plate is explained, together with the three types of plate boundary – ocean ridges, ocean trenches and **transform faults** – and the processes that accompany them. This is followed by a brief discussion of **hot spots** and **plumes**, and finally by a consideration of the driving mechanism of plate motion.

◆ *Chapter 4: Deformation, stress and strain.* This chapter deals with the processes that govern how structures are formed, and with the methods by which the amount of deformation (the **strain**) can be measured. The study of deformation requires an understanding of the relationship between **force** and **stress,** and between stress and **strain**. The various types of strain and the methods of measuring strain are discussed, followed by a consideration of the behaviour of rock material under varying conditions of temperature and pressure in response to stress.

◆ *Chapter 5: Fractures, faults and earthquakes.* The description and geometry of faults and fault systems, and their relationship to their tectonic setting, are described first. This is followed by a discussion of the conditions under which fracturing, and thus faulting, takes place, and the relationship between faulting and earthquakes. Knowledge of how and why earthquakes occur is particularly relevant in today's world – afflicted as it is by so many recent natural disasters with a geological origin.

◆ *Chapter 6: Folds and folding.* This chapter deals first with the description and significance of the various types of fold structure in terms of their shape and mode of formation, and examines the nature and geometry of the strain accompanying each. This is followed by a discussion of fold systems, the three-dimensional geometry of folds, and fold superimposition. **Shear zones** are discussed here as a special case of folding, although they are in reality the deep-seated equivalent of faults. The chapter concludes with a discussion of the conditions under which folding occurs.

◆ *Chapter 7: Fabric.* The effects of higher temperature and pressure are revealed in metamorphic rocks by special types of structure as well as by the changed mineral composition. These new structures include planar structures (**foliation**), new rock types, such as **schist** and **gneiss**, and new linear structures (**lineation**); these are included under the general term '**fabric**'. The effects of deformation under these conditions are to form structures that penetrate throughout the whole rock, unlike folds or faults, and involve micro-structural changes at the scale of the single crystal. Study of the fabric can reveal useful information about the deformation process accompanying the metamorphism.

◆ *Chapter 8: Igneous intrusions.* Igneous intrusions are rock bodies whose shape is controlled by the state of **stress** (or force field) of the crust into which they have been emplaced; the different types of igneous body are thus of structural significance. Moreover, the process of emplacement itself exerts some structural control on the host rocks and produces structures in them. There are important differences between the structures of small-scale bodies such as **dykes** and **sills** and the large bodies (**plutons**) such as **batholiths, laccoliths** and **lopoliths**. Certain plutonic bodies are emplaced passively, by the host rocks moving aside to accommodate them, while others are forcefully emplaced by a mechanism termed **diapirism**, driven by gravitational pressure. Certain igneous complexes representing the roots of large central volcanoes consist of several different types of igneous body (**ring dykes**, **cone sheets** and **radial dykes**) that are related to each other by the stress field generated by the central magma body. Examples of these are described from western Scotland.

◆ *Chapter 9: Structural effects of gravity.* The force of gravity has a key influence on deformation, and this chapter explores how gravity controls structures, giving examples of its effects in terms of gravity sliding on both small and large (**orogenic**) scales, and on the emplacement of **salt domes**.

◆ *Chapter 10: Tectonic interpretation of orogenic belts.* This chapter explains how the principles of tectonics and structural geology are used in understanding present-day and ancient mountain belts.

◆ *Chapter 11: Examples of modern orogenic belts.* Three typical examples are chosen from the present-day orogenic belt system: the central Himalayas, the Western Alps and the Canadian sector of the North American Cordillera. In each case the structure of the belt is explained in terms of its plate-tectonic context, and the sequence of plate movements that are believed to be responsible.

◆ *Chapter 12: Ancient orogenic belts*. The same approach as in the previous chapter is taken to examples of more ancient orogenic belts – the Lower Palaeozoic **Caledonides** of the British Isles, the Mid-Proterozoic **Grenville belt**, and the Early Proterozoic **Trans-Hudson belt** of Canada. Finally, examples of Archaean orogenesis from Canada and Greenland are examined to explore the question of whether plate tectonics and orogeny operated in a basically similar way in the very distant geological past.

◆ *Appendix.* There are certain topics of general geological interest, but not directly relevant to structural geology, which most geology students will understand, but which might be unfamiliar to other readers. These include the geological timescale, and the classification of igneous, sedimentary and metamorphic rocks, which are often useful in understanding the tectonic context of certain structures; these are summarised in the form of tables.

What is not included

This book is not intended as a comprehensive survey of the whole subject, and what is selected partly reflects the author's subjective view of what is most important, and also a desire to keep the book reasonably short. Certain topics are omitted as not directly relevant; these include the *interpretation of geological maps* – the complex geometry of the interaction between simple structures and the three-dimensional ground surface – and *sedimentary structures*, i.e. bedding and allied structures formed by the processes of erosion and sedimentation.

2   Large-scale earth structure

If we were to view the Earth from Space (Figure 2.1), the most obvious large-scale features would be the **continents**, the **oceans**, and the rather linear continental mountain ranges of the Alpine–Himalayan belt and the circum-Pacific belt, which includes the Cordilleran–Andean chain of the western Americas (Figure 2.2). Were we to imagine the water of the oceans removed, another set of mountain ranges would appear in the form of the **ocean ridge** network, which takes up around one-third of the oceanic surface area. **Deep-ocean trenches** would also be apparent, although these are an order of magnitude smaller in areal extent. They form a discontinuous system of linear features parallel to and offshore from island arcs, as in the NW Pacific, or close to the continental margin of western America. These large-scale structures are not randomly arranged but have a tectonic significance, being caused by geological processes that have operated over long periods of geological time.

Mountain belts, ridges and trenches

The most prominent and tectonically active mountain ranges form two linear belts shown in red in Figure 2.2. The Alpine–Himalayan belt extends from Gibraltar, through the Alps, Caucasus, and Zagros ranges, and culminates in the huge mass of the Himalayas and related ranges that sweep around northern India southwards to the Indian Ocean. The eastern side of the circum-Pacific belt chain follows the western margin of the Americas from Alaska down through the North American Cordillera to the Andes of South America. On the western Pacific margin, it forms a discontinuous chain including south-east Siberia, Japan and New Guinea. These belts vary considerably both in width and height: the width of the western American belt is over 600 km in the Canadian Rockies but

Figure 2.1 Earth from space. This photograph shows clearly the continents of Africa, India and much of Asia. The Alpine–Himalayan mountain belt can be faintly seen extending from between the Black and Caspian seas in the top west part of the globe running through the Zagros mountains north of the Persian Gulf and into Pakistan. The Himalayas are lost beneath cloud. NASA photo, courtesy Space Images.

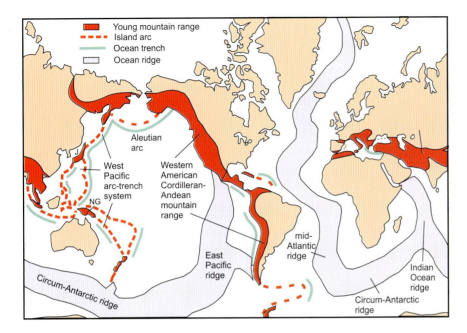

Young mountain range
Island arc
Ocean trench
Ocean ridge

Aleutian arc

West Pacific arc-trench system

NG

Western American Cordilleran-Andean mountain range

East Pacific ridge

mid-Atlantic ridge

Circum-Antarctic ridge

Indian Ocean ridge

Circum-Antarctic ridge

Figure 2.2 Main topographic features of the continents and oceans. The young mountain ranges form two distinct belts, the Alpine–Himalayan and circum-Pacific, meeting in Indonesia. The western part of the circum-Pacific belt consists of partly submerged mountain ranges. The ocean ridges, which are almost entirely submerged, form a continuous network extending through the mid-Atlantic into the Antarctic where it meets the Indian Ocean ridge, and continues into the Pacific, where it ends against the coast of Mexico. The much narrower trenches lie offshore of western America and also parallel to a series of island arcs around the north and west Pacific Ocean and off Indonesia in the Indian Ocean. Smaller arcs are located in the Caribbean and south of S. America.

Earth's internal structure

A considerable amount is known about the internal structure of the Earth, much of it from the study of earthquake waves. On the scale of the whole Earth (Figure 2.3) the **crust** represents only a thin outer shell with an average thickness of around 20 km. The bulk of the Earth is made up of the **mantle,** the base of which lies at a depth of 2900 km below the surface, and is composed

only around 400 km in the southern Andes. The Alpine– Himalayan belt is even more variable: it divides into several branches around the Mediterranean, including the Atlas and Apennine ranges; in central Asia, the belt broadens to around 1500 km where it surrounds the high plateau of Tibet. Ongoing tectonic activity in these belts is expressed by sporadic volcanism and frequent earthquakes. Other prominent mountain belts, such as the Urals, Caledonian and Appalachian ranges, are no longer tectonically active, being the result of processes in more remote geological periods.

The **ocean ridges**, shown in blue on Figure 2.2, are even more prominent features than the mountain belts in terms of their volume, although of course almost entirely hidden below sea level; they are typically 500–1000 km wide and elevated by as much as 2–3 km from the deep ocean floor.

The ridge network consists of three main strands: the Mid-Atlantic, Indian Ocean and circum-Antarctic ridges, the latter extending through the southeast Pacific Ocean to end against the North American coast off Mexico.

The **deep-ocean trenches** (thin green lines on Figure 2.2) are much smaller features than the ridges, being only about 100–150 km across, but they attain depths of up to 11 km below sea level. They fall into two groups: those of the first group lie offshore from, and parallel to, the western continental margin of the Americas, and those of the second group form a discontinuous series of curved features parallel to the volcanic island arcs of the western and northern Pacific Ocean and south-west of Indonesia in the Indian Ocean. Smaller arcs are located in the Caribbean and between South America and Antarctica.

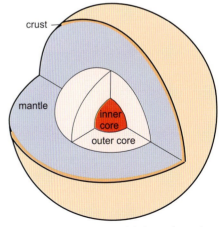

crust

mantle

inner core

outer core

Figure 2.3 The cutaway model shows the main regions of the Earth's interior: solid inner core, liquid outer core, mantle and thin crust.

largely of solid material with a composition similar to the igneous rock **peridotite**. It is here that the processes that mostly control what happens at the Earth's surface originate, as we shall see in the following chapter. The innermost, approximately spherical, region is called the **core**, and extends to the centre of the Earth, at 6500 km depth. The core is believed to consist mostly of metallic iron, with some lighter elements in addition, such as nickel. In the outer core the metal is in a molten state but the inner core is solid. These constituents are thought to have become molten at an early stage in Earth's history and drained down towards the centre, forming the core.

The nature of the crust

The **crust** varies in thickness from about 7 km beneath the oceans, to an average of about 33 km in continental areas, and reaches nearly 80 km in depth beneath certain young mountain belts (*see* Figure 2.4). The composition of the crust is known in some detail, since material from the various depths of the crust is

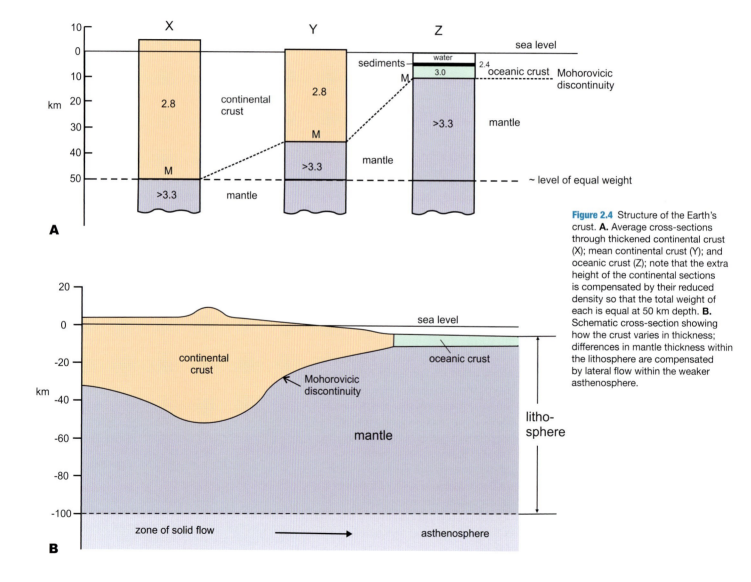

Figure 2.4 Structure of the Earth's crust. **A.** Average cross-sections through thickened continental crust (X); mean continental crust (Y); and oceanic crust (Z); note that the extra height of the continental sections is compensated by their reduced density so that the total weight of each is equal at 50 km depth. **B.** Schematic cross-section showing how the crust varies in thickness; differences in mantle thickness within the lithosphere are compensated by lateral flow within the weaker asthenosphere.

directly accessible at the surface some-where because of tectonic movements. There is a great difference between oceanic and continental crust; **oceanic crust** is composed almost entirely of the volcanic rock basalt, whereas continental rocks are extremely varied in composition, including the whole range of igneous rocks together with all the different types of rock derived from them. However, the average composition has been estimated to be similar to a mixture of granite and basalt.

The present-day coastlines are arbitrary, in the sense that sea level fluctuates through time in response to various geological processes. The formation or removal of ice sheets and the uplift or depression of land masses cause either the retreat or advance of the shorelines across the continental margins. The true margin of the continents may be taken as a line about halfway down the continental slope, where the nature of the underlying crust changes from continental type to oceanic type (Figure 2.5). Measured in this way, the 'continents' would occupy around 40% of the Earth's surface area rather than less than the 30% that currently consists of land. Looking in more detail at the Earth's surface, it becomes apparent that a large proportion of it consists of either plains or plateaux on land, or the deep-ocean floor; these show little variation in relief until interrupted by the extreme elevations and depressions of the mountain ranges and deep sea trenches, which occupy only about 3% of the surface area. Between these two dominant levels is a region of intermediate depths, amounting to perhaps 15% of the total area, representing the continental slopes and the ocean ridges. The reason

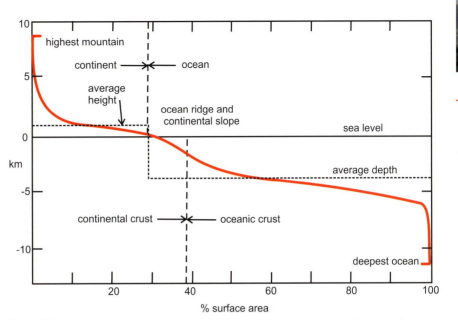

Figure 2.5 The hypsographic curve. This graphical representation shows the variation in surface height (and depth) in terms of the proportion of surface area occupied; there are two large parts of the graph corresponding respectively to continental plains and deep ocean, with a transitional part representing the continental slopes and ocean ridges; areas of extreme relief (mountain ranges and ocean trenches) occupy a very small part of the total.

for this distribution, as we shall see, is that the zones of high relief result from localised tectonic instabilities, whereas the areas of low relief are more stable.

The differences between continental and oceanic areas are due to differences in the composition, and consequently the density, of the underlying crust, as shown in Figure 2.4. **Continental crust**, with a composition corresponding to a mixture of granite and basalt, has a mean density of around 2.8, whereas **oceanic crust**, composed largely of basalt, has a mean density of around 2.9. Moreover, continental crust has a mean thickness of about 33 km and oceanic crust is much thinner, averaging only about 7 km. This explains their difference in mean height (or depth) with

respect to sea level. Because the Earth is in a state of approximate gravitational balance (termed **isostasy**), the weight of any particular sector is similar to that of any other. Consequently, the less dense continents must attain a higher level than the denser oceans if they are to have the same gravitational effect (i.e. have the same weight). In other words, we can imagine the continents as being more 'buoyant' than the oceans, as if they were icebergs floating on the sea. Although the mantle underlying the crust is made up of solid rock (approximating to **peridotite** in composition), it is able to flow in the solid state at a very slow rate of centimetres per year, enabling it to gradually adjust to the pressure of gravitational differences in the crust above. Thus, when a particular part of the crust is subjected to an additional load, by the creation of a mountain range for example, its base adjusts by sinking downwards and displacing mantle material sideways to form a 'mountain root' as shown in Figure 2.4B; like an iceberg, much more of the volume of the thickened crust of a mountain range lies beneath sea level than above! The reverse happens when crust is thinned to form an ocean basin, for example; the top of the mantle rises and the surface is depressed, but the whole is still in gravitational balance.

Asthenosphere and lithosphere

The region of the mantle in which flow takes place lies at a depth of about 100 km beneath the surface, and is termed the **low velocity layer** because of the fact that **seismic** waves travel more slowly through it than would be expected if it were composed of normal mantle material, indicating that its density is lower than that of the surrounding mantle. It is therefore regarded as a zone of relative weakness and is termed the **asthenosphere** (*see* Figure 2.4B). The region above this level, composed partly of mantle and partly of crust, is stronger and is termed the **lithosphere**. As we shall see in the next chapter, it is the lithosphere from which the tectonic plates are formed, and their ability to move across the Earth's surface is due to the relative weakness of the underlying asthenosphere.

Present-day tectonic effects: earthquakes, volcanoes and aseismic movements

The Earth is subject to continuous tectonic activity, as testified by the occurrence of earthquakes (**seismicity**) and vulcanicity of varying severity, much of it with disastrous consequences. This activity is not random but is concentrated in well-defined zones (Figure 2.6).

Earthquakes

Although earthquakes can occur almost anywhere, the great majority of them (and all the most severe ones) are concentrated in narrow zones along the centres of the ocean ridges and in a rather wider zone along the Alpine–Himalayan and circum-Pacific mountain belts delineated in Figure 2.2. The earthquakes along the ocean ridge network are generally quite shallow, less than ~50 km in focal depth, but those near the ocean trenches range from shallow to deep – up to 700 km in depth, indicating that the tectonic process responsible for the latter is different. The earthquakes along the young mountain belts are intermediate between the two extremes, generally lying above about 300 km. The reason for these differences will become apparent when we discuss plate tectonics in the next chapter.

Volcanoes

The distribution of vulcanicity, like that of earthquakes, is distinctly non-random; there are three main categories of occurrence. The majority of active volcanoes are concentrated along the circum-Pacific and Alpine–Himalayan belts, as shown in Figure 2.6; secondly, there are numerous occurrences along the ocean ridges and on subsidiary ridges within the ocean basins, as well as in Hawaii. A third category is

Figure 2.6 Pattern of recent earthquake and volcanic activity. The majority of earthquakes follow well-defined narrow zones along the centres of the ocean ridges and rather broader zones along the Alpine–Himalayan mountain ranges, the island-arc network and the western Americas belt. Most of the volcanoes lie within these same zones, but some oceanic volcanoes are situated away from the ridge crests, especially in the Atlantic ocean.

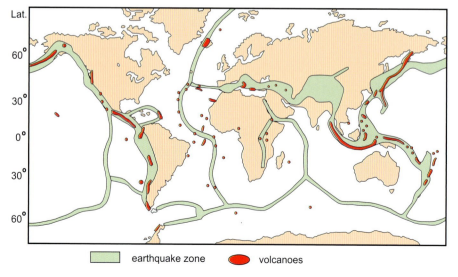

earthquake zone   volcanoes

associated with continental rift zones such as the African rift valley. There are characteristic differences in the type of lava emitted from these volcanoes – oceanic vulcanicity is typically basaltic whereas that associated with the circum-Pacific and Indonesian belts is dominated by **andesitic** vulcanicity. The reasons for both the distribution and the compositional difference are explained by the plate tectonic theory to be discussed in the following chapter.

Aseismic tectonic activity

Crustal movements currently taking place can be measured by repeated recording of location and height using sensitive instruments. The larger and more obvious movements are associated with the major tectonic belts already recognised from their seismic

and volcanic signatures, but less intense movements can be recorded well away from these belts anywhere on the Earth's surface. These movements are generally not accompanied by earthquake activity, that is, they are **aseismic** and show both horizontal and vertical movements of the order of millimetres to centimetres per year.

Relatively recent vertical movements of the land surface relative to sea level can also be studied by observing the positions of **raised beaches** (e.g. Figure 2.7) and buried forests. This evidence shows us that since the last (**Pleistocene**) Ice Age, the land surface of Scotland and Scandinavia has been uplifted due to the slow recovery after the release of the weight of the ice sheet. In the case of Scandinavia, the old shore-line has been uplifted into a dome-like shape by

up to 250 metres (Figure 2.8); this gives average uplift rates of up to 10 millimetres per year. Larger post-glacial uplift rates for Iceland have been calculated in the range 2–9 *centimetres* per year.

Movements of this type characterise broad areas of the crust between the major tectonic belts that may be regarded as stable, in contrast to the instability of the seismic zones. Over long periods of geological time, stable zones (termed **cratons**) can be recognised separating the relatively unstable belts, sometimes known as **'mobile' belts**. Thus, at any given time in the geological past, or at least back to the later **Archaean**, a distinction can be made between cratons and mobile belts that, as we shall see, can be explained by plate tectonic processes.

Figure 2.7 Raised beach. The wave-cut platform in the foreground on which the castle now stands has been raised about 5 m above present-day sea level due to post-glacial uplift; the raised beach is backed by a cliff that would have originally represented the shore line. Lismore, Argyllshire, Scotland; IPR/73-34C British Geological Survey © NERC.

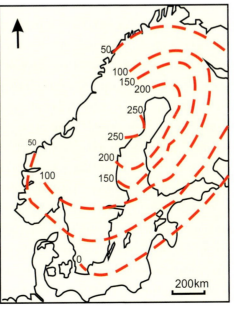

Figure 2.8 Post-glacial uplift of Scandinavia. Contours, in metres, representing the uplift of the Baltic area based on the present-day positions of post-glacial raised beach deposits. The shape of the uplift approximates to an oval dome; the centre of the dome, where the ice sheet would have been thickest, has been uplifted by 250 m. After Zeuner (1958).

3   Plate tectonics

The plate tectonic theory, first advanced in 1967, revolutionised the science of geology in much the same way as atomic theory transformed the science of chemistry at the beginning of the 20th century, providing a new framework into which many previously unrelated or unexplained facts could be brought together and made sense of. This new theory was itself an amalgamation of two previous theories, **continental drift** and **sea-floor spreading**, both of which presented a 'mobilistic' view of the Earth's crust, in which individual continents and pieces of ocean crust were thought to be continually moving around the Earth's surface relative to each other, creating mountain belts in zones of convergence.

Continental drift

Continental drift is a comparatively old idea, first popularised by Alfred Wegener in 1915, which was proposed to explain geometric and geological similarities between continents now separated by oceans. The continents of South America, Africa, India, Australia and Antarctica were shown to fit together in a **supercontinent** called **Gondwana** (Figure 3.1), and North America and Eurasia to fit together into a second supercontinent called **Laurasia**. These two supercontinents appeared to be joined together along the south-eastern coast of North America forming a continuous world-wide landmass termed **Pangaea** (pronounced 'Pan-jee-a')

enclosing the **Tethys Ocean** (Figure 3.2). It should be noted here that, as discussed previously, the term '**continent**' used in a geological sense includes, in addition to the landmass, areas of

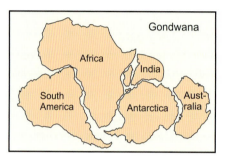

adjacent sea bed underlain by continental-type crust – the **continental shelf** and part of the **continental slope**. When these are included, a much better fit of the Gondwana continents is achieved.

Figure 3.1 Gondwana. This arrangement of the five southern continents 200 million years ago is known as Gondwana. Note how well the coastlines fit together.

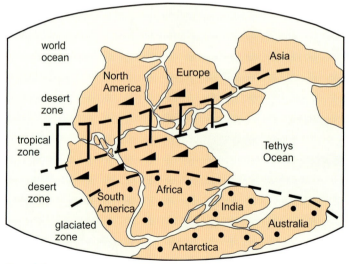

Figure 3.2 Climatic zones of the supercontinent of Pangaea. The climatic zones of 200 million years ago form bands on either side of the equator of that period, which runs through southern North America and Europe, and the polar ice sheet covers large parts of all the present southern continents. Therefore the climatic zones make sense in the Pangaea reconstruction but not when the continents are in their present positions. The tropical zone is defined by coal deposits and coral reefs, the desert zone by dune-bedded sandstones and evaporite deposits, and the glaciated zone by tillites and glacial striations. Based on Hamblin (1989) Figure 17.6, after Wegener (1929).

When this continental reconstruction was examined, many geological features shared by the separated continents could be explained. For example, similar assemblages of fossil land animals and plants that existed over 200 million years ago, prior to the splitting up of the supercontinents, are found in the various separate continents that resulted from the split. The existence of the same land animals and plants in widely separated continents is almost impossible to explain (how did they cross the oceans?) and contrasts with the obvious differences that exist between the assemblages that characterise these continents now.

Another telling piece of evidence is the presence of glacier-derived clays and glacial striations in rocks of late **Palaeozoic** age (~300 Ma old) in all the Gondwana continents, which in their present positions cover about half the globe, but when restored to their presumed Gondwana fit, make a reasonably-sized polar ice cap (Figure 3.2). The distribution of other climatic indicators in these 300–200 Ma-old rocks also makes sense when in the Gondwana fit; these include dune-bedded sandstones and evaporite deposits, which mark out two desert belts on either side of a central equatorial belt indicated by the presence of coal deposits and coral reefs, indicators of tropical conditions. The distribution of these climatic indicator rocks makes no sense in their present locations; for example, coals representing the product of equatorial forests now lie near the North Pole, and glacial deposits lie near the equator! Wegener's ideas caused considerable debate among the geological community, failing to obtain universal

acceptance mainly because of the lack of a convincing mechanism for the movements. Physicists, in particular, opposed continental drift because their calculations of the strength of the Earth's crust 'proved' that it was incapable of the type of behaviour required. However, work on radioactivity led Arthur Holmes in 1931 to demonstrate that the Earth must be much hotter, and therefore much weaker, than previously thought, and suggested that the mantle could be capable of transferring heat by slow flow in the solid state by means of convection currents. Such mantle currents, it was thought, could carry continents laterally across the Earth's surface. Debate carried on, however, until the 1960s when work on **palaeomagnetism** (the magnetic directions of old rocks) showed that the positions of magnetic north for 200 million-year-old rocks in different continents plotted in different places. However, when the continents were fitted together

in their presumed original positions, the locations of magnetic north poles coincided (Figure 3.3). This was convincing proof that the continents had drifted to their present positions from their previous locations in the supercontinent 200 million years ago.

The ocean floor – static or mobile?

The next stage in the evolution of ideas came from studies of the ocean floor, where mapping by various remote-sensing techniques had revealed a topography that was as varied as that of the continents. As described in the previous chapter, the generally even ocean floor is interrupted by a system of great ridges and deep, narrow trenches (Figure 2.2). Much of the objection to Wegener's ideas on continental drift had centred on the failure to visualise how a continent could plough across static ocean crust. However, this objection was countered by the proposal that the ocean

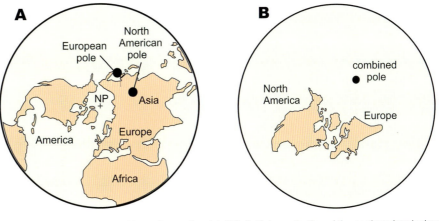

Figure 3.3 Palaeomagnetic evidence for continental drift. **A.** Polar projection of the northern hemisphere showing the different positions of the north magnetic pole for 200-year-old rocks from Europe and North America. **B.** When the continents of North America and Europe are fitted together in the positions occupied in the Pangaea reconstruction, the magnetic north poles coincide. NP, north pole. Based on McElhinny (1973).

crust behaved like a giant conveyor belt, rising at the ridges and moving sideways towards the deep-ocean trenches where it descended (Figure 3.4). In other words, both continents and oceans were mobile rather than static. Wegener had thought that the ocean ridges marked the lines of separation of the continents, but palaeomagnetic dating of the ocean floor of the Atlantic and Indian oceans in the 1960s showed that the ridges were the most recently formed parts, and that the ocean floor became older towards the continental margins.

The dating of the ocean crust relies on the fact that new crust formed along the ocean ridges becomes imprinted with the contemporary magnetic field, and this changes periodically by swapping magnetic north and south poles. Each change creates a long strip of crust, parallel to the ridge axis, whose magnetic character differs from the previous one, and as new strips are created, older strips move away from the ridge axis. This process creates a series of strips (or **magnetic stripes**) on the ocean floor, each representing a particular period of formation (Figure 3.5). The stripe sequence can be calibrated with reference to dated lava flows on land.

The age pattern of the oceanic crust could then be used as evidence that the Atlantic Ocean, for example, has been formed by the creation of new oceanic crust at the mid-Atlantic ridge, as shown in Figure 3.6A. A uniform pattern of ages is evident, becoming younger towards the present ridge axis. The oceanic crust adjacent to those continental margins that had moved apart showed a 'concordant' age pattern, the oldest ages being consistent with the date of separation of the continents.

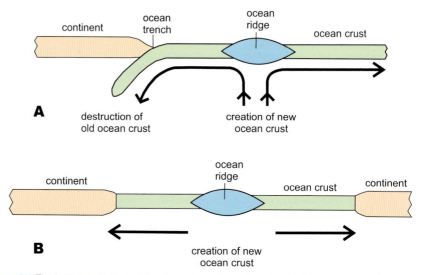

Figure 3.4 The 'conveyor-belt' model. Schematic cross-section showing how ocean crust is created and destroyed: **A.** Creation of new oceanic crust at an ocean ridge and destruction of oceanic crust at an ocean trench; **B.** How continents move apart by creation of oceanic crust.

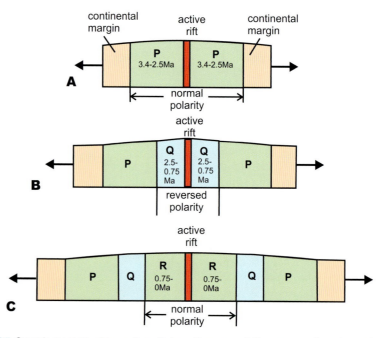

Figure 3.5 Oceanic magnetic stripe pattern. Schematic representative cross-section of oceanic crust resulting from an opening event at 3.4 million years (Ma) ago. New crust (P) is added at the central rift on the ocean ridge from 3.4–2.5 Ma added during mainly normal magnetic polarity; new crust Q added from 2.5–0.75 Ma is added during mainly reverse polarity; and from 0.75 Ma to the present, new crust (R) is added during normal polarity.

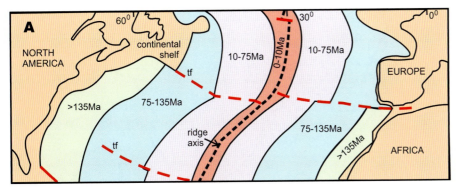

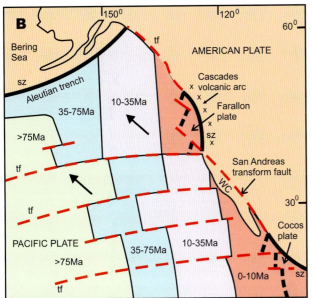

Figure 3.6 Dating the ocean floor. **A.** The age distribution of the ocean floor, dated by magnetic stratigraphy, shows evidence for the opening of the Atlantic Ocean from the concordant pattern of dates, which get younger towards the axis of the mid-Atlantic ridge. **B.** In the NE Pacific Ocean, the age pattern is discordant with the subduction zones and transform faults that make up the northern and eastern boundary of the Pacific plate. The large arrows give the direction of movement of the Pacific plate relative to the American plate. The eastern boundary of the Pacific plate is marked by a ridge (heavy black dashed lines) which separates it from the Farallon plate in the north and the Cocos plate in the south, both of which have been mostly subducted beneath the Americas plate. Transform faults (tf), dashed red lines; subduction zones (sz), thick black lines. Note that part of western California and the Baja California peninsula (WC) is actually part of the Pacific plate. Based on Larson & Pitman (1972).

This palaeomagnetic evidence proved that the continents making up Gondwana and Laurasia had indeed moved apart, and that the space between had been filled by new ocean crust.

However, ocean crust could not be continuously created without it being destroyed elsewhere, and the obvious sites for destruction were the deep-ocean trenches, as suggested in the conveyor belt model. The new palaeomagnetic dating evidence demonstrated that ocean crust adjacent to the trenches shows a variety of ages; that is, the age pattern is 'discordant'. Figure 3.6B shows the age pattern of the north-east Pacific Ocean floor. The age stripes are aligned north–south and become younger towards the American plate, because the oceanic crust of the Pacific plate has been destroyed along the subduction zones on its northern and eastern side.

The above evidence confirmed that the conveyor belt model for the ocean floor was essentially correct.

The plate tectonic model

The final stage in the construction of the plate tectonic model was based on the proposition that both continental and oceanic crust must behave in a semi-rigid manner, moving laterally as single units or blocks, and that relative movement between the blocks was concentrated at their boundaries. This proposition arose from the observation that linear features on the ocean floor, such as faults and the striped magnetic pattern, were essentially unaffected by warping or bending such as might be expected if the ocean floor were to behave in a 'plastic' manner. The opposing coastlines of Africa and South-Central America still show a good fit (*see* Figure 3.1) despite having travelled away from each other for a distance of 3000 kilometres over a period of 150 million years. Thus South America and the western half of the South Atlantic, on the one hand, and Africa and the eastern half of the South Atlantic on the other, could be considered as separate blocks, which moved as units. The Atlantic oceanic age pattern (Figure 3.6A) confirms this.

It was then realised that the linear zones of earthquakes that follow the ridge–trench network (Figure 2.6) must be related to the movements taking place along the boundaries of the relatively stable blocks, and that the margins of these blocks could be mapped out by following the earthquake zones. The term '**tectonic plate**' was introduced to describe these blocks. The earthquake zones completely surround the stable plates, whose boundaries could now be seen to be of three types: **ridges**, **trenches** and **faults** (Figure 3.7). Moreover, since the ridges must mark sites of production of new ocean crust (**constructive boundaries**), and the trenches mark sites of destruction of ocean crust (**destructive boundaries**), the faults must correspond to zones where one block merely slides past its neighbour without either creation or destruction of crust taking place. The faults that form parts of the boundary network link constructive and destructive boundaries and are termed **transform faults**, since they 'transform' one type of motion (e.g. convergent) to another (e.g. divergent). Because plate is 'conserved' at these boundaries, they are known as **conservative boundaries**.

The fact that transform faults, by definition, must be parallel to the direction of relative motion between two adjacent plates can be used to

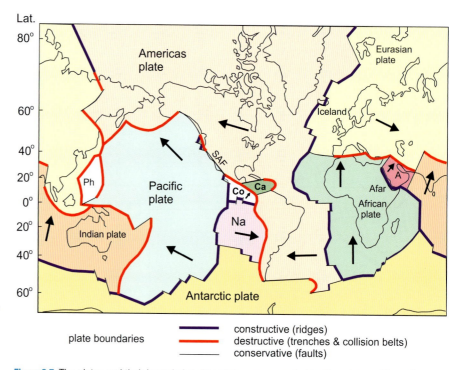

Figure 3.7 The plates and their boundaries. The plates are separated by three types of boundary: constructive – ocean ridges; destructive – ocean trenches; and conservative – faults. Small plates: Na, Nazca; Co, Cocos; Ca, Caribbean; Ph, Philippine; A, Arabian. SAF, San Andreas fault. The arrows give the direction of motion of each plate relative to the Antarctic plate.

determine the relative motion between any two plates that are separated by such a fault. In Figure 3.6A, the orientation of the transform faults tells us that the direction of relative motion between the American and African plates is approximately E–W. Because the plates are, in effect, pieces of a spherical shell (we can imagine them as broken pieces of egg shell) the transform faults must be curved, as shown, for example, on Figure 3.6. In fact, each transform fault represents part of a circle whose centre is the rotation axis of the plate movement, as shown in Figure 3.8.

Figure 3.6B shows a good example of the three types of boundary around

the northern and eastern margins of the Pacific plate. For example, the Pacific plate must be moving in a NNW direction relative to the North American plate, parallel to the San Andreas transform fault, and being subducted along the Aleutian trench in the north. However, the transform faults within the Pacific plate tell us that relative to the East Pacific ridge (of which only part is now visible near the American plate margin) this plate is moving westwards! This only makes sense when we realise that the E–W transform faults indicate only the *relative* motion between the Pacific plate and the Cocos plate, on the eastern side

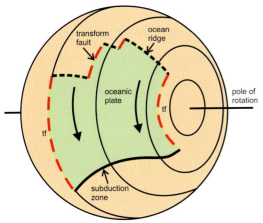

Figure 3.8 Plate motion on a sphere. Lithosphere plates are pieces of a spherical surface; their movement relative to adjoining plates is parallel to transform faults that form part of their common boundary but may be oblique to ridges or subduction/collision zones. The diagram visualises two plates: an oceanic plate (coloured green) is created at a ridge and destroyed at a subduction zone; both these boundaries are connected by transform faults that are parallel to small circles about the pole of rotation of the plate movement relative to the other plate (coloured brown).

of the ridge, which is being subducted north-eastwards beneath the American plate; however, relative to the North American plate, the transform faults themselves are moving northwards.

The rates of movement of the plates can be estimated from the dating of the magnetic sea-floor stripes; these give rates of the order of centimetres per year. Such rates can be verified by present-day precise measurement of the separation of Europe and America, for example. They seem very slow to us but, measured over geological time, are substantial – 100 kilometres in one million years is not untypical. Plate movement at the surface, of course, is not continuous nor uniform but consists of short phases of rapid movement, which generate earthquakes (*see* Chapter 6), separated by long periods of apparent inactivity during which the crust undergoes imperceptibly slow movement known as 'creep'. These movements average out, over geological time, to the 'long-term' rates stated above.

The plates are pieces of the strong upper layer of the Earth, termed the **lithosphere**, consisting of the crust and the uppermost part of the mantle (*see* Figure 2.4B). This strong layer rests on the weaker **asthenosphere**, which is capable of very slow solid-state flow, allowing the plates to move over the underlying mantle. The oceanic lithosphere varies in thickness from around 50 kilometres beneath the ocean ridges to more than 100 kilometres near the continental margins. The continental lithosphere is considerably thicker.

Making new plate: constructive boundaries
The construction of new plate takes place along the central zones of the ocean ridges and within continental rift valleys such as the great **African Rift**. The focus of present-day activity at these constructive boundaries is marked by earthquakes and volcanic activity in a zone around 100 kilometres wide along their central axes.

Figure 3.9A shows how the creation of new oceanic lithosphere takes place. Some of the new material takes the form of basalt magma produced by melting

of the upper mantle in the hot low-density region beneath an ocean ridge. This magma is injected into the crust in the form of basalt dykes or gabbro sills, and part is extruded onto the surface as lava flows. These basaltic rocks form the new oceanic crust; beneath it, new oceanic mantle lithosphere is formed by the addition of ultramafic material transferred by ductile flow from the asthenosphere. As new material is added, earlier formed lithosphere moves sideways, cools, and gradually sinks to the normal level of the deep ocean floor.

The processes attending plate construction at ocean ridges can be conveniently studied in Iceland, which is the only exposed part of the Mid-Atlantic ridge. Here, an average spreading rate of ~2 cm/yr has been achieved by the injection of a dyke swarm along the ridge axis accompanied by a number of volcanoes, two of which (Eyjafjallajökull and Grimsvotn) have been recently active with dramatic consequences. Comparison of precise surface measurements at three of the main volcanic centres by the satellite **InSAR** technique reveals areas of uplift, with rates of ~1 cm/yr, and depression of ~5 cm/yr. The uplifted areas, respectively 40 km and 50 km in diameter, are considered to be due to the accumulation of magma at the crust–mantle boundary, whereas the smaller depressed areas are attributed to subsidence of a shallow magma chamber due to cooling.

The Red Sea–Gulf of Aden–African rift system (Figure 3.9B) is an example of how a constructive plate boundary forms on continental lithosphere. The three rifts form what is known as a **triple junction**; in two of the arms of the system – the Red Sea and Gulf of

A

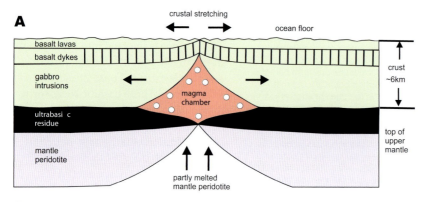

B

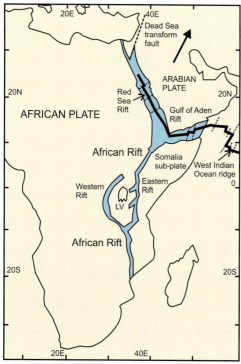

Figure 3.9 Constructive boundaries. **A.** Ocean ridges: creation of new oceanic crust at an ocean ridge. Partly melted hot mantle peridotite rises into the crust to form a magma chamber within the oceanic crust, aided by the stretching of the crust. The magma then separates into an ultrabasic residue, which sinks to the base of the chamber, and a basalt liquid, part of which solidifies within the crust as gabbro intrusions, and part moves up to the surface to form basalt lavas through narrow fissures that become dykes, forming the sheeted dykes layer. As new hot mantle material continues to rise, the newly formed crust and mantle moves to the side and slowly cools to form normal oceanic lithosphere. Based on Bott (1982). **B.** Continental rifts – the Red Sea–Gulf of Aden–African rift system: new oceanic crust is being formed along the Red Sea and Gulf of Aden rifts to join up with the West Indian Ocean ridge, causing the Arabian plate to break away from the African plate. No separation has yet occurred along the African rift, although considerable volcanic activity has occurred here over the last 40 Ma or so. Based on Girdler & Darricott (1972).

of satellite images of surface features before and after a large seismic rifting event in 2005 in the Afar triple junction revealed a horizontal displacement of 6 m associated with faulting, dyke intrusion and volcanic eruption.

The third arm of the system, the African rift, is divided into two branches – the western and eastern rifts. Here the tectonic history is well documented: an initial doming of the surface was followed by collapse of the rifts accompanied by extensive vulcanicity, but there has been no significant separation.

Plate destruction

There are two types of destructive plate boundary (Figure 3.10). The first type follows a zone of destruction of oceanic lithosphere and is marked at the present day by the deep-ocean trenches; the second type is a zone of collision of two continental plates, and is exemplified at present by the Alpine–Himalayan belt of young mountain ranges. This second type of boundary is, in geological terms, only temporary, since convergent movement will eventually cease as the two continental plates grind together and gradually come to a halt.

Subduction

The standard case of the destructive boundary, therefore, is where oceanic lithosphere on one plate (the lower) descends beneath another plate (the upper), which may be either continental or oceanic. The line of descent is marked at the surface by a deep-ocean trench and is known as a **subduction zone**; this zone is typically inclined beneath the upper plate, and its position at depth is indicated by the line of deep earthquakes that follow the inclined plate

Aden – new oceanic crust has formed along the centre of the rifts and joins up with the western Indian Ocean ridge. The separate Arabian plate thus formed is moving north-eastwards away from Africa. Both these rifts

are destined eventually to become oceans as Arabia and Africa move apart. In this way the continents of the Americas, Europe and Africa would have separated during the split-up of Pangaea (*see* Figure 3.2). A comparison

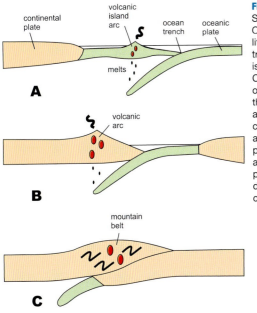

A

B

C

Figure 3.10 Destructive boundaries. Schematic cross-sections showing: **A.** Oceanic subduction zone: cold oceanic lithosphere descends beneath an ocean trench, partly melts, and forms a volcanic island arc on the upper oceanic plate. **B.** Continent-margin subduction zone: the oceanic lithosphere descends beneath the margin of a continent, producing a volcanic arc on continental crust. **C.** continental collision zone: eventually a continent on the lower lithosphere plate reaches the subduction zone, and is underthrust beneath the upper-plate continent; the continental crust is deformed and thickened, and further convergence ceases.

from the remainder of the upper plate to form a new **micro-plate.** This process accounts for the areas of ocean that lie between the island arcs of the western and northern Pacific and the Asian mainland as seen in Figure 2.2.

Plate collision and mountain building
Where two continental plates collide, a zone of crustal overlap is created that leads to a great increase in the thickness of the continental crust. In certain cases, crustal thicknesses of up to 80 kilometres have been recorded, for example in the Alps, compared to a normal crustal thickness of around 33 kilometres. Because of the buoyancy of continental crust compared with the underlying denser mantle, a considerable proportion of this extra crustal material is elevated to form mountain ranges with heights of up to 8 kilometres above sea level; these are supported by a much greater thickness of crustal material in a mountain 'root' beneath

boundary (Figure 3.11). As the lower plate descends into warmer regions of the mantle, some of the crustal material melts, and the resulting magmas ascend into the upper plate, forming igneous intrusions within the upper-plate crust and volcanoes at the surface. Where this zone of volcanoes lies on an oceanic upper plate, it is partly submerged and forms an **island arc**. Present-day examples of volcanic island arcs are widely distributed in the western Pacific Ocean, eastern Indian Ocean, and in the Caribbean (*see* Figure 2.2). Continental-margin subduction zones are situated along the Pacific margins of South and Central America, and of the north-western USA.

Back-arc spreading
Under certain conditions, the subduction process causes the crust on the upper plate, behind the volcanic arc, to be extended and thinned to

form a **back-arc basin** (e.g. see Figure 3.11); in some cases, this basin may be underlain by new oceanic crust, which causes the volcanic arc to separate

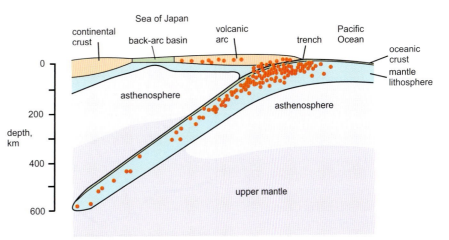

Figure 3.11 Subduction zone beneath the Japan island arc. Cross-section showing the distribution of earthquake foci (red dots) indicating the position of the descending lithosphere slab at depth. Note the back-arc basin between the Japan arc and the mainland. After Uyeda (1971).

(*see* Figure 2.4). This thickened crust consists of rock that has been highly deformed by the effects of the collision; in places great thrust sheets have formed due to the sliding of the upper plate over the lower. The folding and shearing processes caused by the collision (described in Chapters 5–7) are aided by the heating of the rocks by rising magmas formed by melting of the lower crustal material as it becomes depressed into warmer regions at depth. The effects of heat and pressure at depth produce characteristic changes in the rocks and in the types of structure formed, as we shall see.

The best present-day example of such a collision zone is provided by the mountain ranges of southern Asia (e.g. the Himalayan and Pamir ranges) where the Indian continent has collided with, and underthrust, Asia. The record of the gradual convergence of these two continents has been well documented from the ocean floor magnetic data, and there is evidence in the Himalayas of the igneous products of the subduction of the intervening oceanic crust. Therefore, unlike constructive boundaries, which exhibit relatively narrow zones of earthquakes and volcanic activity, such activity in continental collision zones may be over a thousand kilometres wide. The suture surface separating the two opposing plates descends for long distances beneath the surface of the upper plate and may be folded and faulted in a complex manner. The Himalayas are described in more detail in Chapter 11.

The link between mountain belts and plate collision is a valuable tool in interpreting the geological record, especially in old continental crust belonging to periods when no oceanic record has been preserved – all oceanic plate older than about 200 Ma has been subducted. Mountain belts (**orogenic belts**) display characteristic geological features that can easily be identified in the rocks of former periods of Earth history; these are described in more detail in Chapter 10. Such features include:

1. highly disturbed strata (folded and faulted);
2. uplifted metamorphosed deep-crustal material;
3. evidence of enhanced deposition of material derived from the erosion of the uplifted masses;
4. igneous bodies consisting of re-melted crustal material such as granites, rather than mantle-derived basaltic material.

Hot spots and plumes

A map of the Earth's volcanically active regions cannot be simply correlated with the present plate-boundary network. Many such regions are located within the interiors of the plates (e.g. Hawaii and the Azores) and lie on former positions of an ocean ridge. Hawaii is now a long distance from the nearest ocean ridge, but in the case of the Azores, the connection with the mid-Atlantic ridge is more obvious (Figure 3.12). These regions of excessive volcanic activity are known as **hot spots** and in some cases appear to have remained stationary for long periods of geological time relative to fixed Earth co-ordinates. Thus, where plate boundaries have migrated away from them, the

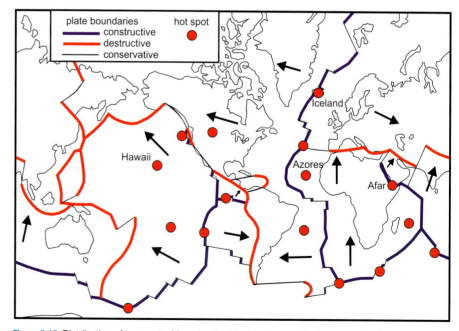

Figure 3.12 Distribution of hot spots. Map showing the locations of the hot spots in relation to the plate boundary network. Note that while some of the hot spots are located on the ocean ridges (e.g. Iceland), several are within oceanic plates, (e.g. Hawaii, Azores). Note also the position of the Afar hot spot on the Red Sea rift.

hot spots now lie well within the plate interiors, as is the case with Hawaii.

On the ridges themselves, activity is not continuous, but is concentrated into more active and less active regions. The best example of an active region is Iceland (Figure 3.13A), where volcanic activity associated with the mid-Atlantic ridge has been so great over the last 50 million years or so that the island has become the only part of the ridge to be elevated above the surface.

Hot spots have been linked to the concept of **mantle plumes** (Figure 3.13B), over which there has been considerable controversy. Some have argued that the plumes represent warm material rising from the base of the mantle, and provide the driving mechanism for plate motion. Sophisticated geochemical techniques have been

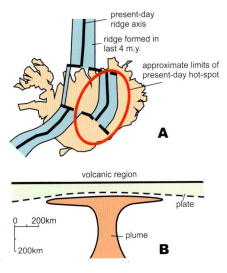

Figure 3.13 A. Simplified map of the Iceland hot spot, believed to overlie a plume. Note that the current position of the hot spot lies east of the ridge, which has migrated westwards away from the hot spot in the last 4 million years. Based on Saemundsson, 1974. **B.** Sketch section of a plume causing a thinning and stretching of the lithosphere above.

devised to categorise the igneous products of plumes, and many examples of excessive volcanic activity recognised in the geological record have been ascribed to plumes on that basis. In the Earth's circulatory system, cooled material is returned to the mantle by means of the subduction zones, whereas the plumes, rather than the ridges, may be the main vehicle for transferring fresh, warm material to the surface. It seems likely, however, that the relationship between convection currents, plumes and plate movements is a complex one for which we do not yet have a satisfactory model.

What drives the plates?

As we have just seen, there is considerable evidence to show how the plates move relative to each other, but why they move is not such an easy question to answer as might at first appear. The early conveyor-belt model for continental drift (*see* Figure 3.4) implied that the movement of the continents was caused by the circulation of mantle material taking place by solid-state flow. This model of the movement of material within the mantle is a modern development of the much older idea of mantle **convection currents**, popularised in the 1930s by Arthur Holmes.

The concept of convection is central to understanding the processes that govern crustal behaviour, and is familiar to all of us from observing the boiling of liquid in a pan heated from below. Warm liquid expands, becomes less dense, and rises to the surface where it is cooled; the heated liquid is replaced by cooler liquid, which, because it is denser, descends to the bottom of the pan.

According to the convection model, the Earth behaves like a giant heat

engine: hot material produced by radioactive decay in the deep mantle rises to the surface and is replaced by descending material supplied by the cooled upper layer. Unlike our boiling liquid analogy, Earth's convection takes place by solid-state flow, which is very much slower than liquid flow, but the experimental study of the behaviour of materials demonstrates that solid state flow can indeed take place, given the right conditions (*see* Chapter 4). The plates, therefore, can be viewed as the cooled surface layer of the Earth's convection system.

Evidence from gravity surveys and earthquake-wave data shows some support for variations in mantle density consistent with the existence of warmer and cooler regions of the deep mantle. However, the detailed geometry of these convection currents has proved difficult to establish, although the simple circulatory cells envisaged originally by Holmes are clearly an over-simplification. In particular, detailed evidence of plate movements demonstrates that the positions of both ridges and subduction zones move laterally across the Earth's surface through time, which is difficult to reconcile with a simple, static, convective-cell system. As an example of this, consider the case of the Antarctic plate, which is surrounded on all sides by constructive boundaries. It follows that the plate must grow in size through time, and that its boundaries must move across the Earth's surface, implying that there cannot be a direct connection to a long-lived uprising hot column. A further problem arises when we consider the positions of the present-day hot spots, such as Hawaii, that are located in the middle of plates.

If, as seems probable, they represent the sites of the upwelling hot currents of the convection system, the lateral flow from them would not always correspond to the direction of motion of the plate, but in some circumstances would oppose it. Another problem is that the part of the upper mantle system in which the most vigorous flow would be expected is the asthenosphere, which, by definition, is the weakest part of the mantle and would seem to be incapable of transmitting the force necessary to move the much stronger lithosphere.

The origin of the forces that drive plate motion, and which therefore are ultimately responsible for crustal structures, must therefore be sought within the lithosphere itself. In the Earth's large-scale convective system that transfers heat, and thus energy, to the surface, heat is lost through the lithosphere mainly at the ocean ridges, and it is at the subduction zones that the cooled lithosphere is returned to the mantle. Both structures are in a state of gravitational imbalance, and it is to this gravitational effect that we must look to explain plate movement. The forces thus generated are called, respectively, the **ridge-push** and **slab-pull** forces (Figure 3.14A).

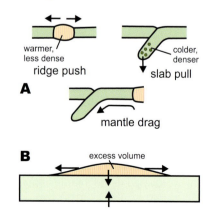

A

B   excess volume

The ridge-push force

An ocean ridge represents a large volume of warmer, lower-density material. The average ridge is around 500 km wide and up to 3 km high, and so represents a considerable additional load on the ocean crust. This extra vertical load is compensated by the lower density of this whole section of lithosphere, in the same way as the mountain ranges are compensated, as explained in Chapter 2 (see Figure 2.4) so that its total weight is the same as that of the surrounding ocean floor. Nevertheless, it is not in gravitational equilibrium, since the gravitational effect of this topographic high is to attempt to restore the extra mass to the general level of the ocean floor, thus providing a lateral force tending to make the extra mass flow sideways (Figure 3.14B). Complete gravitational equilibrium can only be achieved if the various layers of the Earth are of the same thickness throughout and become less dense upwards.

The slab-pull force

The downward movement of a cooled piece of oceanic lithosphere at a subduction zone is due to the gravitational effect of its greater density compared to the adjoining mantle. Its effect is to pull the rest of the oceanic lithosphere to which it is attached along with it, thus providing a sideways force towards the ocean trench; this is known as the **slab-pull force.** This force is opposed by a (usually) smaller force towards the trench exerted on the upper plate of the subduction zone, due to its attachment to the downgoing plate.

Rough calculations of the magnitude of the forces created by these gravitational effects seem to indicate that they are the most likely source of the driving force of the plates. The size of the lateral force provided by material flowing horizontally beneath the plates (**mantle drag** – Figure 3.14A) is an order of magnitude smaller and can probably be discounted. Each of these forces is subject to the effect of frictional forces acting to oppose the plate motion, for example, along transform faults, but these are obviously not large enough to materially affect the movement.

Plate interiors

The forces just described act throughout each plate, as can be shown by the fact that deformation occurs within the plates as well as at their boundaries, although to a much lesser extent. Relatively low-magnitude **intraplate** (i.e. within-plate) earthquakes are widespread and usually caused by movements along pre-existing weaknesses in the crust. The concept of the 'rigid' plate is therefore an approximation, since all plates, particularly continental ones, suffer some internal deformation. In the case of the central Asian section of the Eurasian plate, this internal deformation has been considerable, as we shall see in Chapter 11.

Figure 3.14 A. Mechanisms for plate motion: sketch sections to illustrate: 1) ridge push: warmer, less dense material of the ridge expands and exerts a sideways push on the adjacent plates; 2) slab pull: cooler, denser material of the descending slab exerts a downward pull on the subducting plate; 3) mantle drag: sideways drag of mantle convection current is likely to be much less effective than ridge push or slab pull. **B.** The ridge push mechanism: the gravitational effect of the excess topographic high of the ridge exerts a sideways force on the oceanic plates; the downward gravitational force of this extra mass is balanced by the effect of the less dense mantle root.

4   Deformation, stress and strain

Terms and concepts

Deformation

The term 'deformation' to the structural geologist merely means a change in shape of a rock body – contrary to general usage, there is no implication that this change is for the worse! The study of deformation requires an understanding of the behaviour of rock material under varying conditions of temperature and pressure in response to applied forces. This in turn involves understanding the relationship between **force** and **stress**, and between stress and **strain**. Both the latter terms are used in structural geology, as in mechanical engineering, in a strictly defined way (*see* below) and differently from their everyday usage, where the meanings of stress and strain are more or less interchangeable.

Stress and strain

To the structural geologist, deformation is regarded as the result of a set of forces that act on a body of rock and are converted into a system of stresses (known as a **stress field**); it is these stresses that are in turn responsible for changing the shape of the rock body. This change of shape is known as the **strain**. Strain is therefore a geometrical concept and has to be measured geometrically. To do this accurately, of course, requires detailed measurement and the employment of sometimes complex mathematical models, but often a more qualitative

approach can be just as useful in understanding geological structures.

It is important to realise that it is not possible to work backwards from the final strain to the initial unstrained state or to the causal stress – the relationship must always be based on a guess as to exactly how the strain developed, and there may be an infinite number of possible routes from the initial state to the final result.

Force, stress and pressure

The term '**force**', defined strictly, means '*that which causes acceleration in a body*'; it is the product of the mass of the body and its acceleration. However, this concept is not much use to us in understanding geological structures; with the exception of earthquakes (to be discussed in the following chapter) and volcanic explosions, acceleration is irrelevant to understanding natural deformation in rock. For most practical purposes, the geologist will wish to convert a force, or system of forces, into a stress or stress field. A **stress** is a pair of equal and opposite forces acting on a unit area of a surface (Figure 4.1A) and a **stress field** is the system of stresses acting in three dimensions on a body.

The effect of a force of given magnitude or strength depends on the size of the area on which it acts. This can be

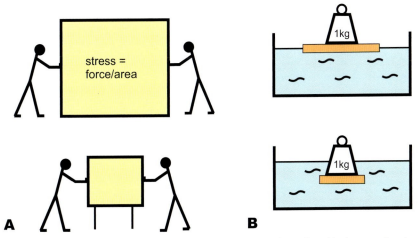

Figure 4.1 Force and stress. **A.** The stress acting on the yellow box is produced by two equal and opposite forces, divided by the area on which the force acts (i.e. the side of the box); a smaller box, subjected to the same forces, will be subject to a larger stress, since the force is divided by a smaller area. **B.** A 1 kg weight resting on a flat slab floats, whereas the same weight resting on a smaller slab sinks, because the force (the 1 kg weight) is spread over a smaller area.

easily demonstrated by placing a weight on a large piece of floating wood, say, which stays afloat, and comparing it with the effect of the same weight on a smaller piece of wood, which sinks under the weight (Figure 4.1B). Thus the same size of force will produce different sizes of stress depending on the surface area of the body on which the force acts, since *stress equals force divided by area*.

Rocks at depth are subject to very great pressure due to the effect of gravity, i.e. the weight of the rocks above. These gravitationally derived stresses are equal in every direction and do not just act downwards, in the same way as water pressure acts at great depth in the sea; this state of stress is termed the **lithostatic pressure** (or just 'the pressure') (Figure 4.2A). In solid rock, this lithostatic pressure is usually added to by directional stresses caused by tectonic forces, which can be either positive (**compressional**) or negative (**extensional**), so that a rock at depth subjected to directional stresses would have two components; the **lithostatic component**, equal to the **mean stress**, plus a directional component (equal to the difference between the maximum and minimum stress) which is the element of the stress field potentially capable of producing deformation (Figure 4.2B).

Normal stress and shear stress
Figures 4.1 and 4.2 show forces, and therefore stresses, acting at right angles to a surface. However, in the general case, a force will act obliquely to a surface (Figure 4.3A). When this force is opposed by an equal and opposite force, the resulting stress can be imagined as being composed of two components: one acting at right angles to the surface, the **normal stress**, and the other acting parallel to the surface, the **shear stress** (Figure 4.3B). The effect of the normal stress is to act across the surface; the effect of the shear stress is to attempt to make the two sides of the surface move in opposite directions. Shear stress is important in understanding the behaviour of faults and shear zones, as we shall see in the following chapters. If we consider the effect of an oblique stress acting across a surface separating two blocks of rock, the size of the shear stress component will influence how readily movement takes place along the surface between the blocks, whereas the normal stress component will either inhibit movement (if compressive) or assist movement (if extensional).

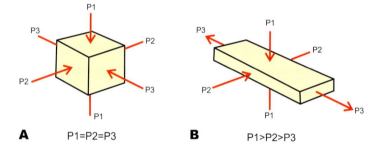

A   P1=P2=P3   **B**   P1>P2>P3

Figure 4.2 Pressure. **A.** in lithostatic pressure, the stresses exerted on a body in every direction are equal; this can be represented by three mutually perpendicular stress axes P1, P2 and P3, where P1=P2=P3. **B.** A variable stress field can be represented by three mutually perpendicular stress axes such that P1 is the maximum stress, P3 the minimum and P2 the intermediate, i.e. P1>P2>P3. The shape of the box illustrates diagrammatically the strain of an initial cube subjected to a stress field of this type. Note that P3, the minimum stress here, is negative, i.e. an extension.

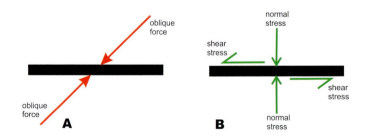

Figure 4.3 Normal stress and shear stress. An oblique force opposed by an equal and opposite force acting on a thin slab (**A**) can be represented by a normal stress acting at right angles to the slab and a shear stress acting parallel to the slab (**B**) and tending to move the top of the slab to the left and the bottom to the right; this is sometimes referred to as a rotational stress; in this case the material inside the slab will tend to rotate in an anticlockwise sense, i.e. the strain will be rotational (see Figures 4.4, 4.5).

Strain
As explained above, strain is defined as the change in shape and/or volume of a body. A change in volume only, termed a **dilation** (Figure 4.4A), which may be

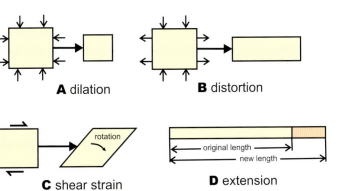

A dilation   **B** distortion

C shear strain   **D** extension

rotation

original length
new length

Figure 4.4 Types of strain. **A.** Dilation = volume change; **B.** Distortion = shape change; **C.** Shear strain: the distortional strain produced by a shear couple; **D.** Strain in one dimension: the extension = the change in length (pink) divided by the original length (yellow).

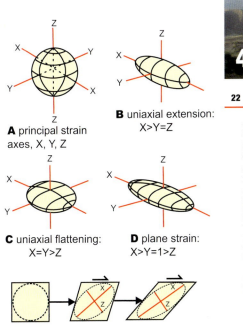

A principal strain axes, X, Y, Z

B uniaxial extension: X>Y=Z

C uniaxial flattening: X=Y>Z

D plane strain: X>Y=1>Z

E progressive shear strain

Figure 4.5 **A–D.** Co-axial strain in three dimensions affecting an initial sphere: **A.** The three principal strain axes X, Y, and Z. **B.** Uniaxial extension: the maximum strain axis (X) is greater than the intermediate (Y) and minimum (Z) axes, which are equal; **C.** Uniaxial flattening: the minimum strain axis (Z) is smaller than the intermediate (Y) and maximum (X) axes, which are equal; **D.** Plane strain: the intermediate strain axis (Y) is unchanged, and all the strain is in the plane of X and Z. **E.** Progressive shear strain: in the XZ plane, the strain axes X and Z progressively rotate with increasing strain.

either an increase or decrease, is difficult to measure in rocks, although we know that volume decreases are associated with large increases in lithostatic pressure by, for example, reducing pore space, expelling fluids and replacing low-density minerals by their high-density equivalents. Changes in shape, on the other hand, lead to characteristic fabrics (rock textures arising from deformation) that are obvious and potentially measurable (*see* Chapter 7). Such changes can consist either of a **distortion** – that is, a shape change (Figure 4.4B), or a rotation, or some combination of both. Distortional strain in response to a shear stress involves rotation of elements of the original unstrained body, and is termed **shear strain** or **rotational strain** (Figure 4.4C).

The description of strain
In one dimension, the amount of strain is measured as an extension, which can be either positive or negative and is the *proportionate amount by which the length of the original body has been changed* (Figure 4.4D).

Thus a body can be said to have been shortened, say, by one-tenth, or 10%. In other words, a strain is just a number, with no attached units.

In three dimensions, the strain in a body whose original shape is known can in theory be measured by the amounts by which the lengths of measured lines in the body have changed. Although the strained body may have been any shape prior to the deformation, it is more convenient in practice to describe the strain as if the original body was equidimensional. The strain can then be described with reference to three mutually perpendicular axes through the centre of the strained body, parallel respectively to the maximum, intermediate and minimum extensions. These axes are termed the **principal strain axes** (Figure 4.5A–D). It is usual to describe the strain as if the original body was a sphere, and the strained body an ellipsoid. Depending on the ratio of the principal strain axes to each other, the **strain ellipsoid** may be either **prolate**, shaped like a rugby ball (Figure 4.5B), **oblate** or pancake-shaped (Figure

4.5C), or what is termed **plane strain**, where the intermediate strain axis is unchanged (Figure 4.5D); or indeed any intermediate shape between these end-member types. Prolate ellipsoids correspond to **extensional deformation**, where the maximum principal strain is appreciably greater than the other two strain axes, which are approximately equal. Oblate ellipsoids correspond to **flattening deformation**, where the minimum principal strain is appreciably smaller than the other two strain axes, which again are approximately equal. In plane strain, all three strain axes are different and the intermediate strain axis is unchanged. This type of strain is characteristic of **shear zones**, which we shall discuss in Chapter 6.

Co-axial and rotational strain

Distortional strain where the strain axes maintain their orientation throughout the course of the deformation is termed **co-axial strain**, as in Figures 4.5B–D. The alternative term for this type of deformation, **pure shear**, is potentially confusing and is not recommended.

Shear strain, on the other hand, **is rotational**, as the strain axes progressively rotate as the deformation continues (Figure 4.5E). This type of strain is also known as **non-co-axial strain** or **simple shear** but the term 'rotational strain' is preferred as being more descriptive.

Since we can only observe the final strain geometry (the so-called **finite strain**) it may not be obvious whether the strain was co-axial or rotational. However, if the unstrained rock contained planar features with varying orientations, the finite strain will differentiate between them as explained below.

Geometrical features of progressive strain

Rocks that contain a variety of planes or lines of varying orientation in their unstrained state, such as many metamorphic rocks, will show progressive geometrical changes as the strain increases in intensity, as shown in Figure 4.6. Depending on their initial orientation, certain planes or lines will extend and others will contract. Moreover, the fields of extension and contraction in the two types of strain differ, depending on whether the strain was co-axial (Figure 4.6A) or rotational (Figure 4.6B). This property is particularly useful in determining the sense

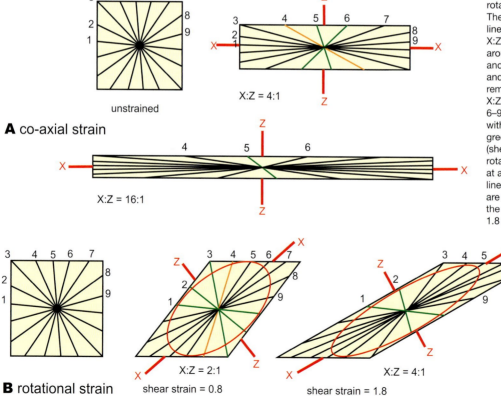

A co-axial strain

B rotational strain

Figure 4.6 Progressive co-axial and rotational strain (in two dimensions). **A.** The effect of co-axial strain on a set of lines 1–9 at 20° intervals: at a strain of X:Z = 4:1 the lines begin to concentrate around the X direction; black lines 1–3 and 7–9 are elongated, green lines 5 and 6 are shortened, and orange line 4 remains the same length; at a strain of X:Z = 16:1 all the black lines 1–4 and 6–9 are elongated and are concentrated within an angle of 6° of the X axis; only green line 5 is shortened. **B.** In rotational (shear) strain, the strain axes (in red) rotate during progressive shear strain: at a shear strain of 0.8 (X:Z = 2:1), black lines 5–9 are elongated, green lines 1–3 are shortened, and orange line 4 remains the same length; at a shear strain of 1.8 (X:Z = 4:1), all of the black lines 3–9 are elongated while green lines 1 and 2 are shortened; line 3, which had shortened at the smaller shear strain, has now elongated.

of rotation (i.e. clockwise or anticlockwise) in shear zones. In Figure 4.6B, the fact that the elongated lines are inclined to the right and the shortened lines to the left, shows that the block has undergone **dextral** (right-lateral) rather than **sinistral** (left-lateral) shear. At large rotational strains, it is possible for planes to first contract and later to extend (as in line 3 in Figure 4.6B), and this history may be revealed by an initially folded layer that has subsequently been pulled apart by later extension.

The measurement of strain
Because rocks are so variable in their physical properties, measuring the strain in any particular part of a complex rock body will not usually be applicable to the behaviour of the whole body. So, rather than attempting to make detailed and accurate measurements of strain in one part of a rock, it will normally be more useful to take a statistical approach and obtain a large number of approximate measurements over a given area. An important restriction is that only certain types of rock will contain objects of known initial shape that can be used to measure the strain; such objects are known as **strain markers**. The best strain markers are initially spherical and include spherulites in lavas, **ooliths** in limestones, and certain reduction spots in slates. The method is illustrated in Figure 4.7A. The assumption has to be made that the strain in the matrix of the rock corresponds to that in the measured objects. This only applies if the two materials have approximately the same strength. Pebbles in conglomerates are frequently used, but allowance has to be made for their initial variability. However, if the initial shape variation is random – that is, there is no original preferred shape orientation – the method can give a reasonable approximation to the overall strain. The shapes of grain aggregates and deformed **phenocrysts** in igneous rocks have also been widely used as strain markers, since at moderate to large strains these objects become ellipsoidal. Certain fossils can also be used, but the geometrical calculations involved can be quite complex and time-consuming.

Another possibility that avoids the necessity of measuring the objects themselves is to measure the spacing between them. If it can be assumed that they were originally spaced either randomly or evenly through the rock body, the spacing between them gives a measure of the strain (Figure 4.7B). The measurement can be simply done by selecting an XZ plane and counting the number of intersections along a traverse in the X direction and comparing it to the number in the Z direction. A similar exercise in either the XY or YZ planes will give the strain ratio X:Y:Z.

A third method can be used where the original unstrained body contains linear or planar elements (e.g. elongate crystals in an igneous rock) that had no initial preferred orientation. This situation corresponds to that shown schematically in Figure 4.6. At a strain of X:Z = 4:1, all but three of the lines are concentrated within an angle of 35° about the X axis, and at a strain of X:Z = 16:1, all but one are within an angle of 22°. Clearly, therefore, at large

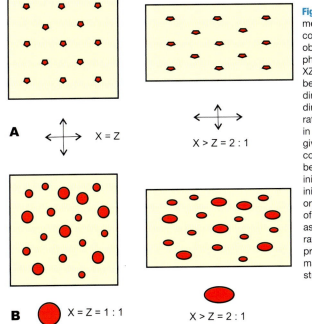

Figure 4.7 Two easy ways to measure strain. **A.** In a rock containing evenly spaced objects (e.g. pebbles or phenocrysts), measured in the XZ plane, the mean separation between the objects in the X direction compared to the Z direction gives the X:Z strain ratio; a similar measurement in either the XY or YZ planes gives X:Y:Z. **B.** In a rock containing objects that can be assumed to have been initially spherical, or whose initial shapes had no preferred orientation, the mean X:Z ratio of the strained objects can be assumed to represent the X:Z ratio of the rock as a whole, provided that the strain in the matrix is equivalent to the strain in the objects.

strains a rough approximation to the strain can be given by measuring the spread of orientations about the X axis.

Structures produced by compression and extension

A body of rock subjected to compression or extension may deform homogeneously if it is composed of material with approximately the same physical properties throughout, as in Figure 4.7, for example. However, if the rock contains layers of stronger material enclosed within weaker, the layers, when compressed, may contract, forming folds (Figure 4.8A). When subjected to extensional strain, a stronger layer may divide into blocks separated by thin necks, or become completely isolated by the enclosing weak material, as illustrated by Figure 4.8B and C. The latter process is termed **boudinage** and the separate blocks, which may be either sausage-shaped or brick-like, are **boudins** (from a French word for a type of sausage).

Transpression and transtension

A block or layer of material may be subjected to compressional stress and shear stress simultaneously, in which case the stress state is termed **transpression**. Shear stress added to extension, is termed **transtension**.

Strain and fabrics

The effect of large strains, mainly on metamorphic rocks but also on some unmetamorphosed rocks, is to produce a set of new microstructural elements that are collectively known as the **fabric**. The subject of rock fabric is discussed in Chapter 7. Fabrics include sets of planar and linear structures (termed respectively **foliations** and **lineations**)

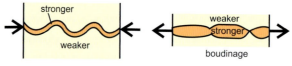

A shortening   **B** extension

Figure 4.8 Contraction and extension in layered rocks. Contraction of a strong layer within weaker material will produce folds (**A**), whereas extension may produce boudinage (**B**). **C.** Extension has resulted in the stronger, sandstone layer being pulled apart forming boudins separated by narrow 'necks'; the weaker shale layers have flowed into the areas between the boudins; see car key for scale.

Figure 4.9 Strain and fabric. Undeformed block (**A**) containing spherical objects is subjected to co-axial strain (**B**): planes parallel to the YZ strain axes are shortened (folded) while planes parallel to the XY strain axis are lengthened and flattened (to give a planar fabric, or foliation). The spheres are extended in the X direction and flattened in the XY plane. A linear fabric, in the form of an elongation lineation is parallel to X.

that penetrate through the whole rock on the scale of the outcrop or the hand specimen, and these fabrics can be used to indicate both the strain axes and, in favourable circumstances, the amount of strain. Thus a foliation that represents a plane of flattening, as in Figure 4.9B, will contain the principal strain axes X and Y, and will be perpendicular to Z. A lineation that represents the direction of elongation within this XY plane will be the principal strain axis X. Under these conditions, therefore, measurements of the strain ratio of flattened or elongated objects in these planes can be used to estimate the strain ratios X:Y:Z.

Behaviour of materials under stress

The behaviour of rock material undergoing deformation is dependent on the inter-relationship between the type of rock (i.e. its **lithology**), embodying its physical and chemical composition, and the physical environment (temperature and pressure) under which the deformation takes place. It is difficult for us to imagine a piece of solid rock being able easily to change shape because it is outside our everyday experience, in the same way that we have difficulty in imagining a solid object consisting mostly of empty space (as the physicists tell us!). However, laboratory experiments help us here in applying stress to material under artificially high temperatures and pressures. Moreover, and critically, by conducting the experiments for long periods of time, more geologically realistic conditions can be approximated, since much of the deformation whose results we can see in the real world has taken thousands or millions of years to complete.

Elastic, plastic and viscous behaviour

The best way to understand how rock material deforms is to first consider the behaviour of familiar materials under stress. Thus the behaviour of a piece of rubber or a spring when extended is said to be **elastic,** and in this type of strain the material returns instantly to its original shape when the stress is removed. Ideal elastic strain can be represented by the amount by which a spring, say, has been extended, and is proportional to the force that is applied to the spring (Figure 4.10A). Rock material can also exhibit an element of elastic strain when a stress is initially applied, although

the nature of the strain quickly changes thereafter if the stress is maintained. In the case of most types of rock material, when a large stress is applied, the material will fracture after only a small amount of elastic strain has occurred.

Most materials, including rock, when subjected to a high enough stress, exhibit permanent deformation – i.e. the strain does not disappear when the stress is removed. This type of deformation is called **plastic strain** and can be simulated by bending a piece of plasticine or putty, or a thin sheet of metal. In the case of the metal, some elastic strain will take place first, but eventually, with an increase in the stress, or even by applying the stress for long enough, plastic strain will occur. Ideal plastic strain can be demonstrated using the analogy of an object, such as a block of wood, being pulled along a rough surface (Figure 4.10B). After an initial force is applied to overcome the friction, steady plastic strain is achieved by applying this force

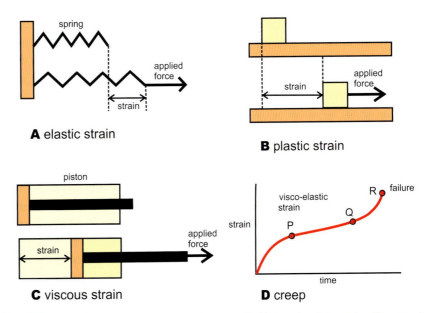

A elastic strain

B plastic strain

C viscous strain

D creep

Figure 4.10 A. Ideal elastic strain is represented by an applied force extending a spring; the amount of strain equals the amount by which the spring has been extended; there is a steady (linear) increase in strain with increase in stress and a steady increase in strain with time at constant stress. **B.** Ideal plastic strain is represented by a block being pulled by an applied force along a rough surface, opposed by sliding friction; the strain is represented by the distance travelled by the block; the frictional resistance represents the yield stress; the strain increases without any increase in stress. **C.** Ideal viscous strain is represented by a piston, drawn through a cylinder by an applied force, the strain being represented by the distance travelled by the piston; the force is opposed by the resistance caused by the viscosity of the fluid in the cylinder; there is a steady (linear) increase of strain-rate with stress and strain with time, at constant stress. **D.** Visco-elastic behaviour and creep: strain-time graph showing a more realistic curve of variation of strain with time, known as a creep curve; initial visco-elastic strain is followed at P by a period of steady-state strain increase (viscous) followed in turn at Q by a period of accelerating strain ending in failure (R).

in pulling the block along the surface at a uniform speed; the amount of strain is represented by the distance travelled by the block. The initial force applied to move the block is analogous to what is known as the **yield strength** of a material – the point at which permanent strain commences. For ideal plastic strain, only this size of force is required; a larger initial force would produce accelerated strain leading to failure.

The concept of **viscosity** is usually applied to the behaviour of liquids although, by analogy, it can be extended to rock materials undergoing solid-state flow. The term **viscosity** is defined as the *rate of flow* of material subjected to a stress; in the case of liquids, it is measured by the rate of flow through a narrow tube, subjected, for example, to gravitational force. Ideal viscous behaviour can be represented using the familiar analogy of the piston, as in the shock absorber of an automobile (Figure 4.10C). Here, the rate of flow (i.e. the *strain rate*) is proportional to the magnitude of the stress; in viscous strain, therefore, it is the *rate of increase* of the strain that is proportional to the size of the stress.

Since viscosity is measured as rate of flow (or strain rate) divided by stress, the larger the value of the viscosity, the more slowly the material deforms. Thus a material with a high viscosity, like rock, deforms more slowly than one of low viscosity, such as oil. The terms plastic and viscous are commonly used interchangeably although, strictly speaking, plastic strain in a given material should exhibit a single strain rate, whereas with viscous strain, a range of strain rates is possible, depending on the size of the stress.

Real rock material deforms usually in a more complex way than these idealised elastic, plastic and viscous models. A closer approximation to the behaviour of real materials is the **visco-elastic** model, which combines an initial period of elastic strain with a period of viscous strain (Figure 4.10D). To simulate the geological conditions under which permanent strain occurs in rocks, such as in the formation of folds, a smaller stress has to be applied for a very long period of time. The resulting variable visco-elastic behaviour is usually known as **creep**. Figure 4.10D shows a typical creep curve of the kind that represents the deformation of real rock materials in laboratory experiments. In this type of behaviour, an initial short period of visco-elastic strain is followed by a much longer period of steady plastic or viscous strain, which may end, after a period of accelerating viscous deformation, in failure. Under geological conditions, the behaviour of rocks will generally fall into two categories: high values of stress lead to an accelerating strain rate and failure after relatively short periods of time, whereas low values of stress lead to long-term, steady, viscous creep at low strain rates.

It is possible to demonstrate creep behaviour of rock in a relatively short time span by suspending a thin slab of rock, such as sandstone, between two supports at each end of the slab. After a period of perhaps months, or even years, depending on the strength of the slab, it will bend downwards and eventually fracture under constant gravitational force. If the amount of downwards displacement (i.e. the strain) is measured over time, it should correspond to a typical creep curve.

Brittle and ductile behaviour

Materials that fail (fracture) after there has been no, or very little, plastic or viscous deformation when a stress is applied are said to be **brittle**, whereas those that experience considerable plastic or viscous flow are said to be **ductile.** These terms are relative and somewhat subjective, and materials that are brittle at low temperatures become ductile at higher temperatures. It follows that, in general terms, brittle behaviour characterises faulting and ductile behaviour, folding.

Effects of temperature, lithostatic pressure and pore-fluid pressure

An increase in temperature has a marked effect on the way a rock deforms, since it decreases the yield strength of the material, so that permanent deformation will commence at a lower stress and the strain rate will increase for a given stress level. Thus an increase in temperature makes a rock more ductile. Increasing the **lithostatic pressure**, however, has the opposite effect, by raising the yield strength of the material, and thus requiring a larger directed stress to achieve permanent deformation.

The effect of raising the lithostatic pressure can be counteracted by the effect of pore fluids. In a rock with a high proportion of pore fluid, the pressure of this fluid (the **hydrostatic pressure**) will approach that of the lithostatic pressure (since the fluid is subject to the same gravitational pressure as the surrounding rock). Thus in rocks at a considerable depth in the crust, the **effective pressure** consists of the lithostatic pressure minus the **pore fluid (hydrostatic) pressure**. This means that, at a given temperature, the ductility of a rock will depend critically on the pore-fluid pressure. This is very important in understanding how large thrust sheets move, as we shall see in the following chapter.

The yield strength of rocks is thus dependent both on temperature and effective pressure, but the effect of temperature becomes more important at greater depth. Consequently, rock in general will become stronger with depth in the crust down to a critical level at which the strength reaches a maximum value, below which it decreases as the temperature increases. This level will be reached at different places with different rocks, but in general will lie at mid-crustal levels. Here there is a transition between broadly brittle behaviour and broadly ductile behaviour in a zone known as the **brittle–ductile transition**, usually regarded as between 15 km and 20 km depth in normal continental crust, below which earthquakes are not usually generated. An alternative way of looking at this is to compare the differential stress required to maintain a geologically realistic strain rate with increasing depth as shown schematically in Figure 4.11. Studies of real rocks in laboratory experiments indicate that the variation in the yield strength is significantly affected by important changes in the mechanism of deformation, which we shall now discuss.

How rocks deform

The various models of deformation discussed above treat rocks as uniform materials in analysing their response to stress. However, to understand their behaviour in practice, it is necessary to examine rocks in detail, on a small scale, especially under the microscope. Since rocks are aggregates of crystals or grains, in most cases involving several different mineral types, their deformation will depend on the various responses of the different minerals to stress. Another important factor is the influence of the temperature–pressure environment during the deformation, since this affects the way in which individual minerals deform. The behaviour of the weakest minerals in a deforming rock is critical, since this controls the strain of the whole rock; in the case of many rocks, this mineral is quartz; in others it is feldspar, or calcite; and in the case of mantle materials, it is olivine; and much experimental work has been done on these minerals. Unfortunately, the short time periods available for experimental deformation means that the strain rates achieved during this deformation are much too high (around 10^{-7} per second, equivalent to 300% per year), whereas we know from plate tectonic velocities that strain rates for lithospheric deformation should be in the range

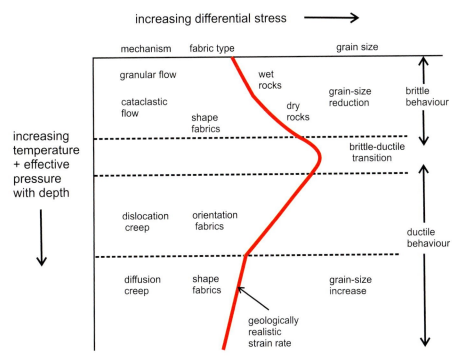

Figure 4.11 Schematic deformation diagram showing the fields occupied by the various types of viscous permanent strain associated with increasing depth (i.e. increasing temperature and effective pressure) and increasing levels of differential stress. The red line represents the behaviour of an imaginary rock material deforming at a geologically realistic strain rate (usually assumed to be around 10^{-14} per sec). At low temperature and high differential stress, viscous deformation takes place by granular flow or, with increasing lithostatic pressure, by cataclastic flow. The strength increases with depth to a maximum at the brittle-ductile transition. Above this level is brittle behaviour, dominated by fracture processes and leading to grain-size reduction. At higher temperatures, lower stresses are required to produce the equivalent strain rate. The mechanisms here are dislocation creep and, at higher temperatures again, diffusion creep. This is ductile behaviour, leading to grain-size increase.

$10^{-12} - 10^{-15}$ per second, or 100,000 to 10 million times slower! Nevertheless, the correspondence between the types of crystal behaviour and rock fabrics seen in these experimentally deformed materials and those characterising natural rock deformation suggests that they do provide a useful guide.

Elastic behaviour

Temporary or elastic strain in a crystalline solid is achieved by distortion of the **crystal lattice**, which is the molecular framework of the crystal (Figure 4.12A). When the stress is removed, the crystal returns to its original shape, but if the stress is maintained for too long, or is increased, the distortion may become permanent and the material begins to behave in a ductile manner.

Permanent strain

Ductile deformation, or **creep**, in rocks is achieved by means of several different mechanisms that produce changes in shape (strain) within the rock that are not recoverable. There are a number of factors that determine which mechanism is chosen in a particular case; these include the nature of the rock (lithology, grain size etc.), the temperature and lithostatic pressure, the pore-fluid pressure, and the differential stress.

Sedimentary rocks near the surface will usually deform by means of **grain boundary** sliding, since the grain boundaries are the weakest part of the rock. In unconsolidated material such as beach sand, for example, this behaviour is obvious, as the grains will merely roll past each other under stress (**granular flow**). In sedimentary material where the grains are cemented together, and in crystalline rocks, small fractures along the grain boundaries will enable the

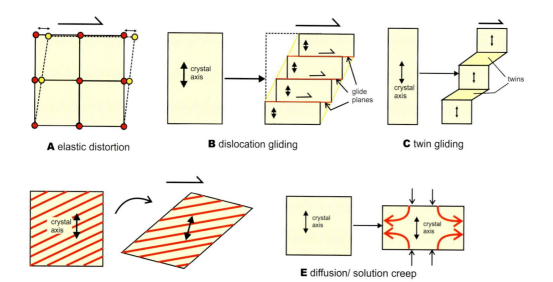

A elastic distortion   **B** dislocation gliding   **C** twin gliding

D rotation + gliding

E diffusion/ solution creep

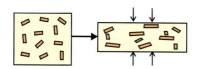

F pressure solution + recrystallisation

Figure 4.12 Deformation mechanisms in crystalline rocks. **A.** Elastic distortion of the crystal lattice. **B.** Gliding on dislocation planes (dislocation creep) within a crystal; the orientation of the crystal axes is unchanged. **C.** Gliding on twin planes within a crystal; the orientation of the crystal axes is unchanged except within the twins. **D.** Gliding on dislocation planes combined with rotation to produce a more favourable alignment with the strain axes; this produces a new orientation of the crystal axes. **E.** With diffusion creep or solution creep, there is mass transfer within the crystal from areas of high stress to areas of low stress; the crystal axes are unchanged. **F.** A combination of pressure solution and recrystallisation causes favourably oriented crystals to expand and unfavourably oriented crystals to contract, since crystals with their strongest axes perpendicular to the stress will be preferentially favoured. Note that mechanisms B–F will all potentially produce a shape orientation fabric, whereas only D and F can produce a crystal orientation fabric.

grains to slide past each other. In this process, the grains will be internally undeformed. Under higher stresses, the grains themselves may be fractured. Creep deformation that is dominated by fracturing is termed **cataclastic flow**. This process produces fabrics characterised by angular-shaped grains and a progressive decrease in grain size. Alignment of elongated grains or grain aggregates may give rise to a **shape fabric**. Frictional heating may give rise to melted crush rock (**pseudo-tachylite**) as described in Chapter 5.

Under higher pressure and temperature, fracturing is replaced by a process termed **dislocation creep**, whereby the crystals themselves are internally deformed (e.g. by **dislocation gliding** or **twin gliding**) so as to achieve a different shape compatible with the overall strain (Figure 4.12B, C). Various features such as **deformation twins** and **kink bands** characterise crystals deformed in this way. Dislocation or twin gliding accompanied by rotation can orient a set of crystals in such a way as to produce the maximum strain effect, and leads to a crystal orientation fabric (Figure 4.12D). These changes in crystals may be accompanied by recrystallisation as the strain progresses, and will be aided by higher temperatures. Crystalline metamorphic rocks deformed in this way will be characterised by a **preferred orientation** of the crystallographic axes, since the newly shaped crystals will tend towards an orientation such that their crystallographic slip planes are aligned favourably with the stress axes. This is termed an **orientation fabric**. Original grains may survive, surrounded by small re-grown grains, producing a characteristic texture.

With further increase in temperature and pressure, another mechanism becomes dominant. Here, material is transferred by diffusion through the crystal lattice or along grain boundaries from areas of high compressive stress to areas of low stress either in the solid state (**diffusion creep**) or by solution and redeposition from a fluid (**pressure solution** or **solution creep**) (Figure 4.12E, F). Ultimately, complete recrystallisation of the rock may take place, leading generally to an increase in grain size. This process will tend to form a shape fabric dominated by elongate crystal shapes that serve to define the new strain geometry; it may or may not produce an orientation of the crystal axes. Solution creep is the typical mechanism in low-grade metamorphic rocks where fluid is available, whereas diffusion creep dominates in higher-grade metamorphic rocks. Figure 4.11 shows schematically how these various mechanisms relate to variations in differential stress and depth for geologically realistic strain rates.

Measurement of present-day crustal stress, strain and strain rate

There are a number of ways of measuring present-day stress in the crust. For example, the analysis of earthquake **fault plane solutions** can yield the stress orientation (*see* Chapter 5) and the use of **strain gauges** can give precise measurements of very small strains and hence, importantly, stress orientations. As would be expected, the state of stress in the crust is closely linked to the plate boundaries, since it is there that the forces that act on the plates largely originate, as explained in the previous chapter. Although complex and variable in detail, on a larger scale, the principal stress orientations are broadly at right angles to convergent and divergent plate boundaries, and oblique at transform faults. Within plates, the stress orientations change gradually from one boundary to the other. For example, the maximum stress orientation in western Europe, which is largely controlled by the mid-Atlantic ridge in the west and the Alpine front in the south, is roughly NW–SE.

The availability of methods of accurate measurement of surface position by satellite since the 1980s has enabled crustal movements, and therefore flow rates, to be calculated with an accuracy measured in millimetres. Using GPS and **InSar** (**Interferometric Short-Aperture Radar**) measurements repeated over time periods of years, displacement rates on faults (*see* Chapter 5) and flow rates in salt and ice (*see* Chapter 9) have been deduced. These give average velocities of several centimetres per year in the case of faults and several millimetres *per day* in the case of salt.

Flow rates for the Tibetan crust resulting from the India–Asia plate collision are in the range 5–15 mm/yr (*see* Chapter 11); these can be used to give an estimate for bulk crustal shortening strain rate of about 10^{-15} (i.e. a shortening of one hundred trillionth!) per second. Taking an estimated shortening of Tibetan crust of ~50% since collision ~50 Ma ago gives the same approximation for strain rate over that period.

5   Fractures, faults and earthquakes

Fractures, faults and joints

Rock fractures are abundant and can be seen at any rock outcrop in the form of cracks interrupting the continuity of the rock surface. Where there has been obvious movement on the crack, the fracture is termed a **fault**; where no displacement is visible, the fracture is termed a **joint**.

Large faults usually result in zones of weakness that are exploited by increased erosion and are marked topographically by gullies, river valleys or escarpments. Figure 5.1A shows the escarpment formed by recent movement on the North Anatolian fault in Turkey. Here the far side of the fault has been raised. In the case of older faults, however, a raised escarpment will usually mark the side of the fault occupied by the harder, or less easily eroded, rocks and will not in general indicate upward movement on the fault.

Describing faults

Faults are classified according to whether the movement occurs up and down the fault plane (**dip-slip** faults) or along the fault plane (**strike-slip** faults) (Figure 5.2). Those where movement is oblique are termed **oblique-slip** faults. Dip-slip faults are normally inclined, with upper and lower sides. The upper side is termed the **hanging-wall** of the fault, and the lower side, the **footwall**. Where the upper block, or hangingwall, has moved downwards on the fault plane, the structure is termed a **normal fault** (Figures 5.1B, C; 5.2A). Where the hangingwall has moved upwards on the fault plane, the structure is termed a **reverse fault** (Figure 5.2B). Normal faults are usually steeply inclined and are much more common than reverse faults.

Measuring fault displacement

The amount of displacement of dip-slip faults is usually measured by the vertical distance, termed the **throw**, by which a given marker horizon has been moved upwards or downwards by the fault, as shown in Figure 5.3A. The angle of inclination (**dip**) of the fault plane should also be recorded, since this is an important indicator of the amount of extension or compression achieved by the fault movement. Measurement of both the throw and the dip will enable the horizontal separation (known as the **heave**) to be calculated. This information can be important in underground surveying, since a borehole sited in the wrong place may either miss an important seam or penetrate it twice (Figure 5.3B)!

A

C

Figure 5.1 Fractures and faults. **A.** The escarpment in the middle distance is formed by uplift of the northern side of the North Anatolian fault, Turkey. **B.** Normal fault cutting bedded rocks; the fault plane dips to the right and the beds in the hangingwall are bent downwards. **C.** Normal fault cutting bedded rocks; the beds in the hangingwall are displaced downwards; note the fractures (small faults and joints in the thick bed that are parallel to the main fault).

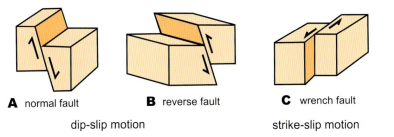

A normal fault   **B** reverse fault   **C** wrench fault

dip-slip motion   strike-slip motion

Figure 5.2 Types of fault. Normal faults (**A**) and reverse faults (**B**) involve dip-slip movement up or down the fault plane (shown in orange); a wrench fault (**C**) involves strike-slip movement (along the fault plane).

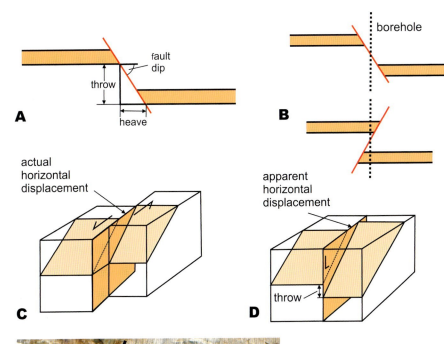

In the case of strike-slip faults, it is the horizontal displacement that is important, and here it is ideally necessary to find a vertical marker that has been offset in order to calculate the displacement. Merely using an inclined plane, such as bedding, to measure horizontal displacement is misleading, since horizontal offset can just as easily be caused by a vertical movement, as indicated in Figures 5.3C and D. The displacement on oblique-slip faults can be described in terms of dip-slip and strike-slip components.

Movement along fault planes often produces a polished surface, which is termed a **slickenside,** and the striations or grooves scored on this surface by the grinding movement of irregularities on the opposite fault wall, termed **slickenlines,** indicate the movement direction on the fault plane. Another type of linear structure is caused by elongate crystal fibres (**slickenfibres**) that have grown parallel to the movement direction on the fault plane (*see* Figures 5.3E and 7.6B, C). These elongated crystals, usually of quartz or calcite, grow from little step-like irregularities on the fault wall and point in the direction of movement of the opposite fault wall. This type of fault plane structure is useful in distinguishing normal from reverse faults and sinistral from dextral strike-slip faults.

Figure 5.3 Measurement of fault displacement. **A.** Throw (vertical displacement) and heave (horizontal displacement) in a normal fault. **B.** The heave in a normal fault causes a gap in the faulted bed, whereas the heave in a reverse fault causes an overlap of the bed. **C.** The horizontal displacement on a fault can be measured by the displacement of a dipping layer only if the fault displacement is entirely horizontal (i.e. strike-slip). **D.** An apparent horizontal displacement of an inclined layer may be caused by a vertical displacement only. **E.** The linear structure on this fault surface is caused by elongated calcite crystals (slickenfibres) that have grown parallel to the direction of fault movement (see black lines). Note that the direction of movement in the lower half of the picture is different and represents a separate (earlier or later) stage of faulting.

Fault systems
Extensional fault systems

Sets of normal faults such as those shown in Figure 5.4 are typical of regions where the Earth's crust is being extended and thinned. Such regions occur, for example, at **constructive plate boundaries** – that is, along the **ocean ridges** and **continental rift** systems. Extensional fault systems also occur locally in mountain belts due to lateral spreading of the thickened crust – a process known as gravitational collapse (*see* Chapter 10).

Normal faults are typically arranged such that uplifted and depressed blocks are bounded by sets of parallel or sub-parallel steeply dipping faults, inclined away from the uplifted blocks (known as **horsts**) and towards the depressed blocks (**graben**) as in Figure 5.4A. However, only a limited amount of extension can be achieved by this arrangement, and for further extension to take place, the faults need to be rotated into a more gently inclined attitude, as seen in Figure 5.4C. This has the result of also rotating the fault-bounded blocks, so that originally horizontal strata within the blocks also become inclined.

Gravitational effects of large extensional faults

Large extensional displacements of the crust result in local thinning, which in turn reduces the weight of that particular section of the crust. To maintain gravitational equilibrium, it is necessary for the underlying mantle material to rise so that the total weight of that section remains the same. Thus in Figure 5.4C, the mantle material beneath the detachment fault has bulged upwards to balance the effect of the thinned crustal material. The same process occurs in large normal faults, as shown in Figure 5.4B. Here, the gravitational response of the extensional crustal thinning is to produce an **antiformal fold** in the hangingwall (often termed a 'rollover') and a **synformal fold** in the footwall. In this way, the gravitational imbalance that

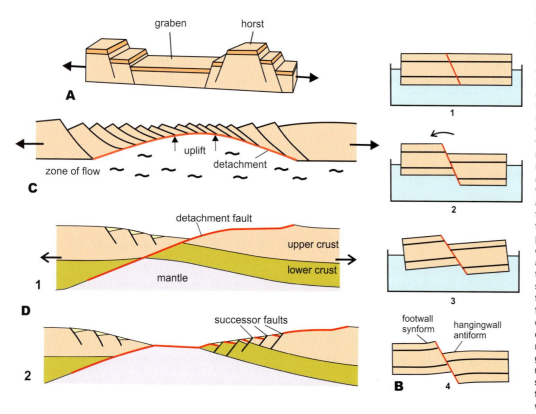

Figure 5.4 **A.** Extension on a set of steeply-dipping normal faults creates down-faulted blocks (termed graben) and up-faulted blocks (termed horsts); the orange layer has been extended by a relatively small amount, equivalent to the cumulative heave on the normal faults. **B.** The gravitational effect of a large normal fault can be visualised by considering the effect of a wooden block floating in a tank of water (1); once faulted (2), the block is unstable and must tilt (3) to maintain equilibrium; (4) the tilted sections of the block are represented in nature by a hangingwall antiform (rollover) and a footwall synform. **C.** Cross-section through the crust showing how larger amounts of extension cause initially planar faults to become curved and rotate due to solid-state flow in the lower crust. This process results in extension and thinning of the upper crust, leading ultimately to arching up of the lower-crustal material (see text for further explanation). **D.** Cartoon sections (after Reston, 2007) showing how the crust may be stretched and thinned to form an oceanic basin by the development of a low-angle normal fault that acts as a detachment horizon; 1, as the hangingwall moves to the left, the footwall rises to retain gravitational equilibrium; 2, the mantle, roofed by the detachment, is now at the surface; successor normal faults now cut through the footwall into the lower crust, which is now brittle.

would have resulted if no bending had taken place is smoothed out.

In some cases, normal faults, as they descend into lower levels of the crust, may become curved; such faults are termed **listric faults** and may even become sub-horizontal, allowing the fault blocks above them to rotate and slide over the basal fault, which acts as a detachment horizon (Figure 5.4C). This allows much larger extensions to take place. Figure 5.4D shows how stretching of the crust by means of a low-angle normal fault may ultimately lead to the rise of the mantle to the surface and the creation of an ocean basin. As the structure evolves, the original detachment fault becomes inoperative in the footwall and further extension takes place along steeper 'successor' normal faults.

Extensional faulting in petroleum exploration

Many of the world's oil and gas fields are hosted in structural traps created by extensional faulting (e.g. Figure 5.5B), and much of the present-day petroleum exploration has focused on the mapping by remote means of under-sea extensional structures. Such structures are typical of extensional basins and passive continental margins (e.g. the North Sea and the Atlantic margin west of Shetland, where exploration activity in the UK is currently concentrated). Our detailed knowledge of extensional faulting is largely due to interpretative **seismic sections** produced in the course of petroleum exploration.

Figure 5.5A shows a typical example from the western North Sea basin of the structural and stratigraphic information that can be obtained by this method.

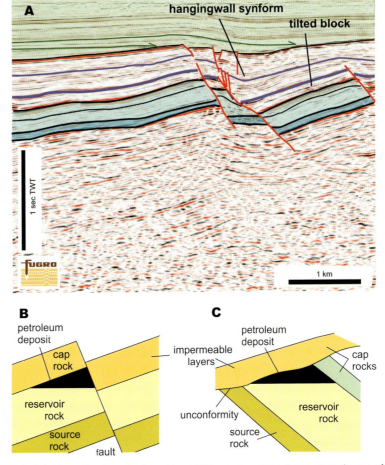

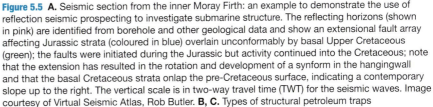

Figure 5.5 A. Seismic section from the inner Moray Firth: an example to demonstrate the use of reflection seismic prospecting to investigate submarine structure. The reflecting horizons (shown in pink) are identified from borehole and other geological data and show an extensional fault array affecting Jurassic strata (coloured in blue) overlain unconformably by basal Upper Cretaceous (green); the faults were initiated during the Jurassic but activity continued into the Cretaceous; note that the extension has resulted in the rotation and development of a synform in the hangingwall and that the basal Cretaceous strata onlap the pre-Cretaceous surface, indicating a contemporary slope up to the right. The vertical scale is in two-way travel time (TWT) for the seismic waves. Image courtesy of Virtual Seismic Atlas, Rob Butler. **B, C.** Types of structural petroleum traps

The North Sea basin is dominated by a series of grabens developed during the Triassic to Jurassic period and the petroleum deposits, which mostly originate in Jurassic shales, are trapped in the overlying sandstone formations and capped by impervious rocks such as the Upper Cretaceous Chalk Formation (Figure 5.5B, C). Determining the detailed geometry of the structural traps is therefore critical in finding and exploiting petroleum reserves.

The stratigraphy in Figure 5.5A can be calibrated from borehole data, and shows a set of normal faults of Upper Jurassic age that have continued to be active, though with smaller throws, into the Cretaceous, displacing the basal Upper Cretaceous unconformity.

Formation of a passive continental margin

The role of extensional faulting is critical in the initiation and evolution of a constructive plate boundary, and the subsequent formation of an ocean basin. Since much of the evidence for this is necessarily below sea level, studies of passive continental margins have largely relied upon seismic surveying, aided by deep-ocean drilling and the use of submersible dives. Such methods, employed in the investigation of the (non-volcanic) west Iberian rifted continental margin has revealed a complex array of extensional faults belonging to several generations, each of which has become inactive (locked) and subsequently replaced by a younger, steeper set (Figure 5.6).

As indicated in Figure 5.4, the gravitational effect of the extension is to elevate the mantle together with the earlier formed faults, which rotate into a sub-horizontal attitude. Ultimately, as shown in Figure 5.6A, the oldest and lowest fault, now a basal crustal detachment, becomes back-tilted, and mantle material is exposed at the surface.

Compressional fault systems

A special category of reverse fault, termed a **thrust fault,** is typically gently inclined or even horizontal. Major thrusts occur in zones in which numerous individual thrusts and reverse faults are linked together in a complex manner, all contributing to the overall movement. In bedded rocks, thrusts may describe a **staircase** path by alternately following a weak layer, often a bedding plane, known as a **detachment horizon**, then rising up a **ramp** to achieve a higher level (Figure 5.7A). Progressive movement of the upper thrust sheet causes these ramps to be transported across flat-lying strata to form **fold** structures

(Figure 5.7B) in the overlying sheet. These fault-generated folds are discussed further in the following chapter.

Eventually, the gravitational load on a rising thrust sheet causes the thrust plane to stick – or lock up – in which case, continuing compression will cause a new thrust to develop. This second thrust usually forms in advance of the original ramp by extending the basal thrust along the detachment horizon, then cutting upwards to form a new ramp. A succession of such structures can result in a very complex assemblage, as shown, for example, in Figure 5.7C.

Thrust faults play an important role in mountain belts by transferring rocks from deep crustal levels up to the surface, moving them over younger, originally higher-level, rocks. In this way they respond to crustal shortening by increasing crustal thickness. Present-day mountain belts such as the Alps, the Himalayas and the North American Rockies (*see* Chapter 11) all exhibit sets of thrusts that have undergone displacements of many tens or even hundreds of kilometres. A famous example of a much older thrust system is the **Moine thrust zone** of north-west Scotland (Figure 5.7C), which has been known and studied since the 1880s – *see* Chapter 12. It is the marginal regions

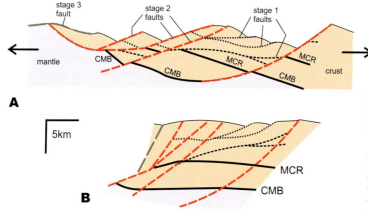

A

B

5km

stage 3 fault

stage 2 faults

stage 1 faults

mantle

CMB

CMB

MCR

MCR

CMB

crust

MCR

CMB

Figure 5.6 Structure of the west Iberian rifted continental margin. **A.** This seismic section, constrained by several deep-sea drill cores, reveals a complex array of extensional faults belonging to several generations, each of which has become inactive (locked) and subsequently replaced by a younger, steeper set; the gravitational effect of the extension is to elevate the mantle together with the earlier formed faults, which rotate into a sub-horizontal attitude, ultimately becoming back-tilted, and exposing mantle material at the surface. **B.** A partial restoration shows the crustal section before fault movements on the (red) second-stage faults. CMB, crust–mantle boundary; MCR, mid-crustal reflector. A, B, based on Reston (2007).

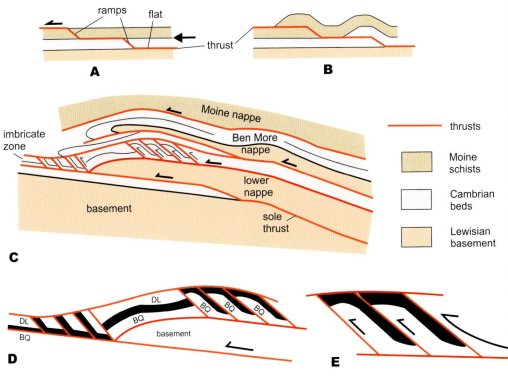

A

B

ramps flat

thrust

C

Moine nappe

imbricate zone

Ben More nappe

lower nappe

sole thrust

basement

thrusts

Moine schists

Cambrian beds

Lewisian basement

D

DL

DL

BQ

BQ

BQ

BQ

BQ

basement

E

Figure 5.7 Thrust faults in profile. **A.** Thrusts in layered rocks usually follow bedding planes (flats) for part of their course, then cut up through the beds on a ramp structure. **B.** The result of this flat/ramp geometry is that the thrust sheet forms folds to accommodate to it; synforms are caused by ramps in the footwall, and antiforms where hangingwall ramps overlie a flat. **C.** Simplified diagrammatic W–E cross-section through part of the Moine thrust zone of NW Scotland, showing how successive thrust sheets (termed nappes) bring older, deeper, rocks from further east upwards towards the west. One of the nappes is made up of slices of bedded Cambrian rocks bounded by steep reverse faults; this is termed an imbricate structure (**D**). **E.** Enlarged part of the imbricate nappe showing how each slice is formed by bedding being pushed along the basal thrust and up the steep reverse faults. BQ, basal quartzite; DL, Durness limestone.

of mountain belts that typically display major thrust zones of this type, known as foreland fold-thrust belts. Such zones generally follow a pattern in which the outer thrust sheets are formed from bedded rocks of the **foreland** (the region outside the mountain belt), whereas further inwards towards the interior of the belt, they are replaced by much larger thrust sheets derived from deeper levels of the crust, as in Figure 5.7C. The latter are often quite different from the foreland rocks and include metamorphic rocks strongly altered by heat and high pressure, since the thrusts have reached down towards the base of the crust. It is because of this that we are able to study rocks formed at depths of up to 80 kilometres beneath the surface.

Strike-slip fault systems

Strike-slip faults (also known as **wrench faults**) are typically vertical or steeply inclined and separate blocks of crust that have moved past each other in opposite directions in a sub-horizontal direction (see Figure 5.3C). Sets of such faults are typical of conservative plate boundaries, of which perhaps the best-known example is the **San Andreas fault** in western USA (Figure 5.8A – see also Figure 3.6). This structure is actually a fault zone up to 100 kilometres wide containing many individual faults that form a branching network. Because only a short section of a particular fault is active at any given time, the opposite movements of the blocks on either side create zones of compression or extension at the ends of the moving blocks. Subsidiary structures such as normal faults may be formed in the extended zones and reverse faults in the compressed zones. Another complication may arise at bends in the main fault. At such bends, the fault is oblique to the main direction of movement, creating zones of compression or extension respectively (Figure 5.8B, C). Active or recent strike-slip fault zones such as the San Andreas are typified

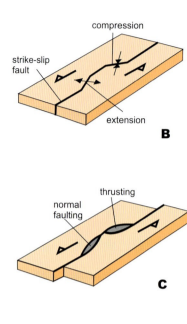

Figure 5.8 The San Andreas fault. **A.** Aerial view of a section of the San Andreas fault showing how clearly the fault is reflected in the topography by a valley along the line of the fault and uplifted zones on each side. Photo © Jim Wark. **B, C.** Bends in the course of a strike-slip fault create zones of extension and compression resulting, respectively, in subsidiary normal faulting and thrusting along those sectors of the fault that are oblique to the direction of movement. B, before movement; C, after movement.

At greater depth, fault rocks are much harder and more cohesive. Here, frictional heating caused by rapid movement on the fault plane may cause local melting of the crushed rock, forming a material termed **pseudotachylite,** which forms veins intruding into the surrounding rock (Figure 5.9B). Like the igneous basaltic rock, **tachylite**, the matrix is a glass, in this case formed from the melted crush rock. Fault rocks at even greater depth may become partly recrystallised to form a metamorphic rock. A fine-grained banded variety of such a rock is termed **mylonite** (*see* Figure 7.5C) and is found along major deep-level thrusts such as the Moine thrust of NW Scotland. At mid-crustal levels and lower, faults and fault rocks are replaced by **shear zones**, which are characterised by intense deformation and recrystallisation, as discussed in the following chapter.

Physical conditions for fracturing

The process of fracturing, as discussed in the previous chapter, is the result of brittle behaviour where the stress, and therefore the strain rate, is higher than the level required to maintain continuous viscous flow in the deforming material (*see* Figure 4.11). These conditions normally apply only in the upper 10–15 km of the crust, although fracturing also occurs in subduction and collision zones where cooler rocks have been transported to greater depths. As explained in the previous chapter, fracturing is critically dependent on several variables: notably, the magnitude of the applied stress, the temperature, and the effective pressure (lithostatic pressure minus pore-fluid pressure), in addition to the strength

by a close association of uplifted blocks that have been created in the compressional zones and depressed basins in the extensional zones.

Strike-slip faults forming part of the plate boundary network, such as the San Andreas fault, are known as **transform faults** and are described in Chapter 3. Active transform faults in the oceans are very common and are marked by periodic shallow earthquakes. Underwater surveying by side-scan sonar has identified linear ridges and troughs along the course of some of these faults.

Fault rocks

Movement on a fault produces a zone of broken and crushed rock fragments of varying size; this process is known as **cataclasis**. When formed near the surface, **cataclastic rocks** may be composed of large angular fragments and are termed **fault breccia** (similar in appearance to sedimentary breccia)**,** as shown in Figure 5.9A; where the fragments are small, a type of clay is formed, called **fault gouge**. Such rocks are soft and easily eroded, which explains the fact that many faults are followed by valleys.

Figure 5.9 A. Fault breccia with a fine-grained matrix showing a slight foliation. **B.** Black pseudotachylite vein containing small rock fragments intruding cataclastic gneiss (x5).

the former controls the displacement along the fracture plane, whereas the latter acts across the fracture plane and will either aid or inhibit displacement.

Relationship of shear fractures to the stress axes

The planes of maximum shear stress intersect along the intermediate principal stress axis and theoretically lie at 45° to the maximum and minimum stress axes, as shown in Figure 5.10A. In real materials, however, the shear fracture planes always make a smaller angle (a) than 45° to the maximum stress axis, and this angle varies both with the type of material and with the lithostatic pressure.

Strength of materials

The measurement of strength is obviously of critical importance in

of the rock itself. Moreover, when considering the conditions for faulting to take place, it is necessary to take into account both the **shear stress** and the **normal stress** components on a given fracture plane (*see* Figure 4.3), since

assessing the suitability of materials in manufacturing and construction. It is of equal importance in rocks, in determining whether they are strong enough to support buildings or dams, for example. A given material possesses three critical values of strength – **compressive strength**, **tensile strength** and **shear strength**, any or all of which may be important depending on the use to which it is being put. Moreover, the shear strength depends also on the angle of the plane of shear to the direction of the applied stress, as shown in Figure 5.10A.

Materials can be tested in the laboratory under varying conditions of applied stress and lithostatic pressure to find the range of critical values at which the material fails. The complex relationship between normal stress, shear stress and lithostatic pressure is essentially a three-dimensional problem, but can be reduced to two dimensions by assuming that the intermediate principal stress is approximately equal to the mean stress. The stress state at failure (Figure 5.10B) is controlled by the *normal stress* (which we can assume equals the effective lithostatic pressure) and the *shear stress* (which is proportional to the *stress difference*). Using this simplification, the stress conditions at failure for a given material at constant temperature can be represented by the **shear failure envelope** (Figure 5.10C) where both the shear stress and normal stress at failure are represented by a point on the envelope. The gradient of the envelope, represented by the angle 'β' in Figure 5.10C, is related to the angle α that the shear fracture makes with the principal stress axis (β = 90°-2α); i.e. as β decreases, α increases towards

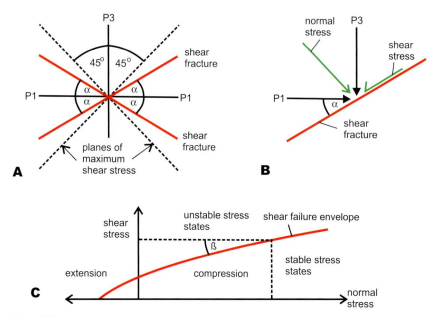

Figure 5.10 Stress conditions for shear fracture. **A.** as seen in 2D, shear fractures (red) intersect on intermediate principal stress axis P2 and make an angle α with the maximum principal stress axis P1; this angle is smaller than the angles of 45° which the planes of maximum shear stress make with the P1 and P3 axes. **B.** This diagram illustrates the relationship in two dimensions between shear stress, normal stress and the two principal stresses, P1 and P3 for a fracture plane making an angle a with the direction of maximum compressive stress, P1. Note that the shear stress will be proportional to the stress difference (P1 – P3) and that the normal stress will be proportional to the mean stress (P1 + P3)/2. **C.** The shear failure envelope, represented by the red line, shows the variation in normal stress and shear stress at failure for a given material. Note that the shear strength is much smaller under extension than under compression, and that the strength increases with increasing normal stress. The angle β (= 90° – 2α) which measures the gradient of the failure envelope also decreases with increasing normal stress, signifying an increase in the angle α towards its maximum value of 45°.

its maximum theoretical value of 45°. Where the shear stress decreases to zero, at the left end of the graph, failure is by extensional fracture, at right angles to the maximum principal stress. The effect of pore-fluid pressure is to reduce the effective normal stress component. It is clear from this diagram, based on a typical rock material, that:

1. failure occurs at a lower shear stress under extension than under compression – i.e. its tensile strength is less than its compressive strength;

2. the shear stress required for failure (its *shear strength*) increases with increasing lithostatic pressure;

3. the strength under extension (its *tensile strength*) is limited to the point at the left end of the failure curve, but there is no theoretical limit to the compressive strength if the shear stress lies within the failure envelope. This illustrates that materials in general are much stronger under compression than under tension.

The angle α that the shear fracture plane makes with the maximum compressive stress decreases with increasing normal stress; i.e. with increasing lithostatic pressure.

The shape of the shear failure envelope is specific to a particular material; thus in laboratory experiments, for example, it was found that the curve for dolerite was steeper than that for marble; that is, the shear strength of the dolerite increased more rapidly with increasing compressive stress.

Application to fault orientation

The above relationships can be used to predict the geometry of shear faults under idealised conditions in homogeneous rocks (Figure 5.11). Because there are three possible orientations of the three principal stress axes, and we can assume that shear stress parallel to the Earth's surface must be zero, two of these principal stress axes must be horizontal. This leads to three theoretical types of fault arrangement.

◆ Where the maximum principal stress is vertical (usually equivalent to gravitational load), two sets of normal faults can form, with opposed dips (Figure 5.11A), especially where the minimum principal stress is negative (i.e. extensional).

◆ Where the maximum principal stress is horizontal and the minimum principal stress vertical, two sets of thrust faults are predicted, again with opposed dips (Figure 5.11B); this situation is more likely to apply at relatively high levels in the crust, where the effect of gravitational load is less important.

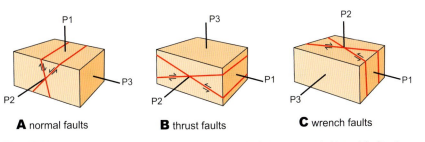

A normal faults   **B** thrust faults   **C** wrench faults

Figure 5.11 Orientation of shear faults in relation to principal stress axes. **A.** Normal fault set: maximum principal stress P1 vertical, minimum principal stress P3 horizontal. **B.** Thrust fault set: maximum principal stress P1 horizontal, minimum principal stress P3 vertical. **C.** Wrench fault set: maximum and minimum principal stresses, P1 and P3 both horizontal.

◆ Where both the maximum and minimum principal stresses are horizontal, two sets of wrench faults can form, their orientations bisected by the maximum stress axis (Figure 5.11C).

Only the normal fault set-up is common in nature, and is typical of regions having experienced crustal extension. Thrust sets are usually highly asymmetric, and appear to be driven by an overall shear sense in a particular direction; that is, they are subject to kinematic, or movement-based, control rather than dynamic, or stress-based control. Wrench fault sets are similarly asymmetric, being often parallel to large regional transform faults. However, the dynamic model of shear-fault orientation is much more useful in interpreting joint patterns, as explained below.

Rates of displacement on active faults
Studies using precise ground positioning by GPS and InSAR methods have established rates of displacement along a number of important, presently active, strike-slip faults. In several of the major active faults in Tibet, such as the **Altyn Tagh** and **Karakorum** faults,

displacement rates typically vary in the range 5–15 mm/yr (*see* Chapter 12). Observations over many decades on certain sectors of the San Andreas fault in the western USA, which is a plate boundary, reveal a pattern of rather higher average slip rates of 3–4 cm/yr made up of long periods of quiescence interrupted by seismic events resulting from displacements of the order of several metres, indicating that the crust experiences a slow build-up of strain over long periods of time before being suddenly relieved by rapid movements generating an earthquake.

Joints
All rock outcrops display numerous cracks or small fractures that generally lack obvious signs of displacement and are too small in scale to be classified as faults. Such cracks are termed **joints**. Joints typically form sets with similar orientations and origins, although in many cases they are irregular and unsystematic. Because erosion preferentially attacks the joints, the rock outcrop may become carved into a series of blocks defined by the joint surfaces (e.g. *see* Figure 5.12A).

Joint sets
The most common type of joint is extensional in origin, and is formed either by contraction of the rock, or because it has been subjected to an extensional stress. Such joints are often filled by vein material such as quartz or calcite that has been deposited by percolating solutions. Under ideal conditions, sets of extensional joints will be perpendicular to the direction of extension of the rock, that is, the minimum principal stress, but often the orientation of such joints is very variable. In sedimentary rocks, joint sets are often both parallel to bedding and at right angles to it (**cross joints**). Where the bedding is folded, sets of cross joints may occur parallel to the fold axes (especially along the fold crests) and also at right angles to the fold axes; the latter are termed **longitudinal joints** and indicate that some extension has taken place parallel to the fold axis.

Another type of joint is a shear fracture formed under compression, similar to a small-scale wrench fault. Pairs of such shear fractures may form in such a way that the smaller angle between them is bisected by the compression direction. Sets of shear joints thus mimic the idealised fault sets illustrated in Figure 5.11 and may comprise both thrust-sense and strike-slip sets. Such joint sets are useful in identifying the orientation of the stress field acting on the rock.

It is important when interpreting joints to recognise that early-formed joints in bedded rocks will become re-oriented during folding. For example, cross-joints that were originally parallel may become more radial and open up on fold crests, as seen in Figure 6.2B.

Moreover, later stress conditions may cause originally extensional joints to become shear fractures and *vice versa*.

Unroofing joints

Another variety of extensional joint is formed as a result of what is termed **unroofing**. This occurs when rock above the present ground level is stripped off by erosion, thus reducing the gravitational pressure on the rock below. The release of this pressure causes the rock to expand, typically forming sets of joints parallel to the bedding in horizontal stratified rocks, or sub-parallel to the present ground surface in igneous rocks, as shown by the example in Figure 5.12A.

Cooling joints

Contractional joints are associated particularly with igneous bodies, where they have formed as the rock contracts on cooling. In some cases, these joints form a set of polygonal columns at right angles to the cooling surface. The Giant's Causeway in County Antrim, Northern Ireland, and the island of Staffa, west of Mull in north-west Scotland, are two well-known examples of this phenomenon. The columns of Staffa, formed within a lava flow (Figure 5.12B), are vertical and have hexagonal cross-sections, like a honeycomb.

Initiation and propagation of a fault

Faults are caused by sudden brittle rock failure after a period of increasing strain in the zone around the site of the future fracture. Initial elastic strain is followed by limited permanent strain caused by the opening of small cracks in the strained zone as

Figure 5.12 Joints. **A.** Joint sets in granite tor, Dartmoor, Devon: the main joint set (due to 'unroofing') is sub-horizontal and roughly parallel to the ground surface. **B.** Columnar jointing in a horizontal basalt flow, Staffa, north-west Scotland; note the vertical joints arranged in a hexagonal pattern and also the prominent horizontal joints.

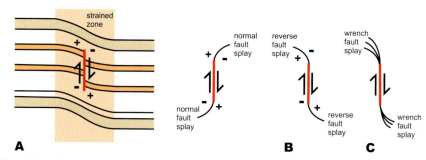

Figure 5.13 Initiation of a fault. **A.** A strike-slip fault develops within a strained zone, causing the formation of compressional (+) and extensional (-) areas at the ends of the active fracture. **B.** Splay faults may develop at the ends of the active fracture: normal fault splays in the extensional zones and reverse fault splays in the compressional zones; alternatively, (**C**) some of the displacement may be taken up by wrench fault splays.

the shear strength of the host rock is approached. These preliminary strain effects cause zones of limited local expansion in the rock within the strained zone, which can be monitored to give some indication of impending failure. When failure does occur, the elastic element of the strain is suddenly released causing an earthquake (Figure 5.13A). The release of strain around the new fracture causes the stress to increase at the ends of the new fracture causing an increase of strain there, and leading eventually to an extension of

the fracture. As Figure 5.13A shows, the stress fields at the ends of the fracture are changed as a result of the shear displacement because one side of the fracture will experience compressional stress and the other extensional. This has the effect of encouraging new fractures to form at an angle to the existing fracture plane; these are termed **splay faults** (Figure 5.13B, C).

If the initial failure occurs within an existing fault zone, it may spread rapidly along it over distances of many tens of kilometres. The major San

Andreas fault zone of California has experienced countless earthquakes along its 1200 kilometre length over a period of several million years, contributing to a total displacement estimated at several hundred kilometres.

Earthquakes

An **earthquake** is a vibration of the ground caused by a sudden displacement or failure of rock at depth. The vibrations caused by this event spread out in all directions from the source of the disturbance, like ripples in a pool after a stone has been thrown in. Earthquakes are common and very widespread in occurrence; most are too small to be detected except by sensitive instruments, but the largest ones cause immense damage and loss of life.

Intensity and magnitude: the 'size' of an earthquake

The destructive power of an earthquake – its **intensity** – depends on the severity of the ground motion, and is measured by the effects felt at the surface. Intensity is usually described in terms of numbers on the **Mercalli scale** (Table 5.1) and ranges from instrumental (detected only by instruments) to catastrophic (total destruction of all buildings). In populated areas, it is possible to map zones of increasing intensity towards a central position above the source of the earthquake where the intensity is at a maximum. This point is termed the **epicentre**. The **magnitude** of an earthquake is a more exact way of describing its size, and is usually measured on the **Richter scale** (Table 5.1). The magnitude measures the amount of

energy released by the earthquake at its source and is reflected in the severity of the surface vibrations, the '**earthquake waves**'. The most severe earthquakes known have a magnitude of between 8 and 9 on the Richter scale, whereas the weakest detectable by humans have a magnitude of about 3.5. Those weaker than this are only detectable by instruments. Each successive number on the Richter scale represents a factor of ten times more energy than the last; thus a magnitude 5 earthquake is ten times more severe than a magnitude 4, and so on.

The damage and loss of life caused by a major earthquake can be catastrophic, as demonstrated by several notable examples in recent years. Two contrasting cases are the Haiti earthquake in January 2010 and the more recent devastating earthquake in Japan in March 2011. The magnitude-9 Japanese earthquake was caused by a deep-seated movement on the subduction zone along the

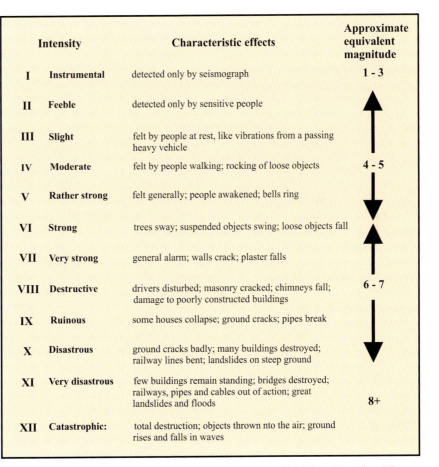

	Intensity	Characteristic effects	Approximate equivalent magnitude
I	Instrumental	detected only by seismograph	1 - 3
II	Feeble	detected only by sensitive people	
III	Slight	felt by people at rest, like vibrations from a passing heavy vehicle	
IV	Moderate	felt by people walking; rocking of loose objects	4 - 5
V	Rather strong	felt generally; people awakened; bells ring	
VI	Strong	trees sway; suspended objects swing; loose objects fall	
VII	Very strong	general alarm; walls crack; plaster falls	
VIII	Destructive	drivers disturbed; masonry cracked; chimneys fall; damage to poorly constructed buildings	6 - 7
IX	Ruinous	some houses collapse; ground cracks; pipes break	
X	Disastrous	ground cracks badly; many buildings destroyed; railway lines bent; landslides on steep ground	
XI	Very disastrous	few buildings remain standing; bridges destroyed; railways, pipes and cables out of action; great landslides and floods	8+
XII	Catastrophic:	total destruction; objects thrown nto the air; ground rises and falls in waves	

Table 5.1 Characteristic effects of earthquakes of different intensities (Mercalli scale) and the approximate equivalent magnitudes (Richter scale).

western margin of the Pacific plate, and the resulting displacement of the ocean floor produced a huge ocean wave, or tsunami, that devastated vast areas of low-lying ground. In this example, the destructive effects were due much more to the tsunami than to the earthquake vibrations themselves. In the 7.0 magnitude Haiti earthquake, the focus lay at a shallow depth of only 13 km a mere 25 km west of the capital, Port au Prince. It caused widespread destruction; an estimated 3 million people were affected and tens of thousands lost their lives. The cause was movement on a transform fault separating the eastward-moving Caribbean plate from the westward-moving Americas plate (see Figure 3.7).

In general, destructive effects bear no simple relationship to the size of an earthquake and depend much more on the quality and construction methods of the buildings involved, and indeed whether the earthquake occurs in an inhabited region. In recent years severe earthquakes have occurred in Japan with much less consequential damage because the buildings there are constructed to withstand earthquake vibrations.

The causes of earthquakes

There are over one million earthquakes every year, but very few of these are large. Their frequency is inversely proportional to their magnitude: thus in a given year there might be about 700,000 earthquakes of magnitude 1 but only 20 of magnitude 8. The amount of energy released every year by earthquake activity is probably approximately constant. This energy results from the continual pressure exerted by the forces acting on the Earth's moving plates (see Figure 3.14) and is released when the pressure builds up to the point when it overcomes the strength of the rocks. This happens more often at plate boundaries, where relatwive plate motion is accompanied by large opposing forces, but because the resulting stresses are transmitted through the plate interiors as well, any weakness, such as an old fault, can result in the release of part of this stress in the form of an intraplate (within-plate) earthquake.

Earthquakes originate within those zones of the Earth's crust and uppermost mantle where the rock is strongest and most liable to sudden failure. In the lower crust and in the greater part of the mantle, the rocks are weak and tend to flow rather than fracture. According to their depth of origin, earthquakes are divided into shallow focus (depths down to 70 km), intermediate (from 70 to 300 km) and deep (between 300 and 700 km). Shallow earthquakes are typical of ocean ridges and faults along constructive or conservative plate boundaries, as explained in Chapter 2 (see Figure 2.6). Intermediate and deep earthquakes, on the other hand, are confined to destructive plate boundaries, where cooler and therefore stronger slabs of crust and uppermost mantle have been thrust down to much greater depths (see Figure 3.11). Earthquakes are heavily concentrated along plate boundaries, and help to define them; however, numerous shallow earthquakes also occur within plate interiors, which are also subjected to stresses, albeit of lesser magnitude than those concentrated at the plate boundaries.

Although earthquakes are usually associated with the formation or re-activation of faults, they can also occur as a result of volcanic activity, particularly when accompanied by the explosive release of gas.

Earthquake waves

The shock waves from an earthquake are recorded on an instrument called a seismograph, on which the ground vibrations are traced on a revolving drum in the form of a series of wave forms (Figure 5.14A). The first waves to arrive are termed primary (P) waves, and are followed by a second set, termed secondary (S) waves, followed in turn by a third, termed surface waves. Both primary and secondary sets take a 'short cut' through the Earth, whereas the surface waves travel around the Earth's surface and thus take longer to arrive. The arrival time of the surface waves is directly related to the distance from the earthquake source measured around the surface. It is the surface waves that are responsible for most of the earthquake damage.

The location of the source of an earthquake may be determined using the time difference between the primary and secondary waves, which increases with distance from the source. A minimum of three seismographs situated in different directions from the source are necessary to find the epicentre, which is located at the intersection of the three circles with centres at each of the three seismograph stations and radii equal to their respective distances from the epicentre (Figure 5.14B).

By comparing seismograph records of the first motion of larger earthquakes in various parts of the globe, it is possible to determine the sense and

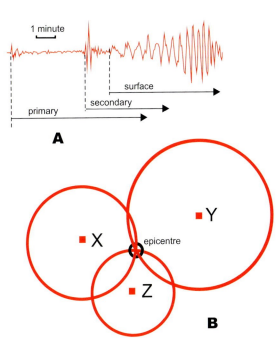

Figure 5.14 Earthquake waves. **A.** Typical earthquake-wave recording (seismograph), showing primary, secondary and surface wave traces. **B.** Method of locating the epicentre of an earthquake detected at recording stations X, Y and Z; the radii of the circles represent the respective distances from these stations to the epicentre.

zones such as the **San Andreas fault zone** of Western USA, which marks a plate boundary (*see* Figure 3.6), repeated earthquake events have occurred in the past and can confidently be predicted in the future, for as long as there is relative motion between the Americas plate and the Pacific plate to the west. Sections of the fault zone that have recently experienced movement are unlikely to fail again soon, but may transmit stress to adjoining sections. The longer a specific segment of the fault zone is dormant, the larger the earthquake event is likely to be when it occurs.

It is possible to monitor sections of a fault zone that lie within or near inhabited areas to detect the small changes in the ground that precede failure. As explained above, the rocks contract or expand slightly due to increases or decreases in stress. Pores and cracks may open up, affecting the flow of fluids, and these changes are accompanied by changes in physical properties, such as electrical resistance. However, the exact time of the earthquake is impossible to predict with sufficient accuracy as yet, and of course failure might come in a part of the zone that has not been monitored.

direction of movement on the fault; this is termed the **'fault-plane solution'**. Unfortunately there are two possible solutions in each case, so that it is necessary to make an assumption about which of two possible orientations of the fault is the correct one. From this data it is possible, for example, to distinguish between dip-slip and strike-slip motion and to determine which of several possible fault sources is responsible for a given earthquake.

Earthquake prediction

Much research has been carried out in an attempt to accurately predict major earthquakes. This has been only partially successful. Along large fault

6   Folds and folding

Folds are beautiful and intriguing structures that attract the interest and curiosity of all geologists – which of us has not paused on a mountain path to admire a particularly fine example? They come in a wide variety of sizes and shapes. The largest are many kilometres across and can only be viewed from the air or on a map (Figure 6.1); the smallest are microscopic. There is also a bewildering range of shapes – from rounded to angular, regular to irregular, and so on. These geometric characteristics are referred to by structural geologists as the fold 'style' and reflect differences in lithology (i.e. the physical properties of the folded rock layer), and the physical conditions (temperature and pressure) under which the folding took place. The fold style can vary markedly between one rock layer and the next, and even within the same layer – look at the amazing variety of fold shapes in Figure 6.2. Before examining these shape variations, however, and the reasons for them, it is necessary first to establish a method for describing folds.

Describing the fold shape
The parts of a fold

The simplest fold shape consists of two **fold limbs** separated by a **hinge** – the line marking the maximum change in orientation of the folded layer (Figure 6.3A). The **fold hinge** (sometimes referred to as the **fold closure**) may be sharply defined, with relatively straight limbs, or curved, approaching the surface of a cylinder (compare Figures 6.2B and C), in which case the hinge is a zone rather than a well-defined line. The **fold angle**, as measured by the smaller angle between the two limbs (Figure 6.3B) can be used to give a rough minimum estimate of the amount of shortening achieved by the folding. Most geologists generally only distinguish between **open folds** (fold angle greater than ~90°) and **tight folds** (less than ~90°) although some recognise further categories; **isoclinal folds** are a special case of tight folds where the limbs are parallel. The size of a fold can be described in terms of its width (or **wavelength**), the distance between two adjacent anticlinal (or synclinal) hinges, and its height (or **amplitude**) (Figure 6.3C).

Figure 6.1 Large-scale folds from the air, Bighorn County, Wyoming. The prominent anticline in the centre right of the picture plunges gently away from us. To its right is another anticline and to its left, in the distance, is an anticlinal pericline. Between these anticlinal structures there are synclinal areas with less regular outcrop shapes. © Jim Wark.

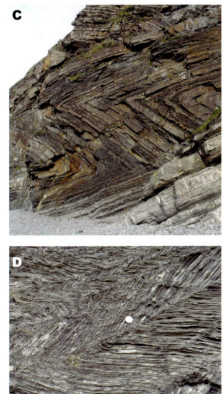

A

B

C

D

Figure 6.2 Some typical fold styles. **A.** Folds with variable profile shape in interlayered slates and sandstones; the sandstone beds show near-parallel geometry, whereas the slates show chevron geometry. **B.** Buckle folded sandstone beds with near-parallel geometry; note the extension fractures around the hinges. **C.** chevron folding in thin sandstone beds with slate interlayers; the cliff face is about 20 m in height. **D.** Close-up of chevron folding in thinly layered slates; note the straight limbs and sharp hinges; 50p coin for scale. **E.** Similar folding; note how the profile shape is maintained through many layers. **F.** Disharmonic folding: the wavelength of the upper parallel-folded thick layers is much longer than that of the thinly-layered material below.

Fold orientation

The orientation of a fold is usually recorded by the attitude of its hinge and axial plane. The **axial plane** (Figure 6.3B,C) is an imaginary plane at equal distance from each limb, and can be described by its trend (**strike**) and inclination (**dip**). The line along which the axial plane intersects the fold surface is known as the **fold axis**, and is a more geometrically precise measure than the hinge (Figure 6.3C). A more useful method for describing folds where the axial planes of the

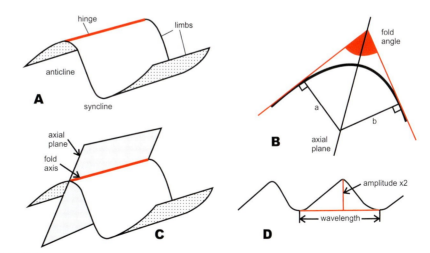

Figure 6.3 Fold description. **A.** folds consist of two limbs separated by a hinge. **B, C**, axial plane, fold axis and fold angle; the axial plane is equidistant from the limbs (a = b). **D.** wavelength (width) and amplitude; note that the amplitude = half the height.

different layers do not correspond, such as the example in Figure 6.2B, is to use the **axial surface**, which is the surface containing the hinge lines of each successive folded layer. Recording the orientation of groups of folds in this way can yield information about the direction of the compressive stresses responsible for the folding.

Types of fold

Folds with near-vertical axial planes are known as **upright**, as in Figure 6.4A. In folds with inclined axial planes, one of the limbs may have rotated through the vertical to become inverted; such folds are termed **overfolds**, and in the extreme case, where the axial plane is horizontal, they are termed **recumbent** (Figure 6.4B). The inclination of the hinge line of a fold, i.e. the angle between the fold axis (or hinge) and the horizontal, is known as the **fold plunge**.

In a set of folds, each individual fold shares its limbs with the adjoining folds on either side (Figure 6.4A). Upwardly convex folds are termed **antiforms** and downwardly convex folds, **synforms**. These terms are to be preferred over the more familiar terms, 'anticline' and 'syncline'. This is because, strictly defined, **anticlines** should have older rocks in the core of the fold, and **synclines** younger; in the usual sequence of bedded rocks, which get younger upwards, antiforms would be anticlines and synforms, synclines; but where the folded layer has been inverted, in a region of very complex folding, antiforms may actually be synclines, and synforms, anticlines (Figure 6.4B)! However, in the great majority of cases, antiforms probably are anticlines and vice versa.

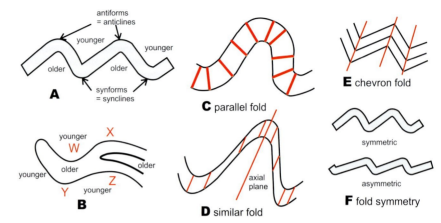

Figure 6.4 Types of fold in profile. **A.** Antiforms, anticlines, synforms and synclines: antiforms close upwards and synforms close downwards; anticlines have older rocks in their core, synclines have younger, so antiforms = anticlines and synforms = synclines only where the beds are right way up. **B.** In a refolded recumbent anticline, W is both a synform and a syncline; X is both an antiform and an anticline; Y is an anticlinal synform and Z is a synclinal antiform. **C.** Parallel fold: layer thickness is constant and the fold surfaces are parallel to each other. **D.** Similar fold: the distance across the layer measured parallel to the axial plane is constant; each fold surface has the same profile shape. **E.** Chevron fold: the limbs are straight and the hinges sharp and angular. **F.** Symmetric folds have the same limb length; asymmetric folds have unequal limb lengths.

The shape of a fold, or set of folds, is usually described by viewing the fold in profile, perpendicular to its axis, as in Figure 6.4. Single-layer folds can be described as rounded or angular depending on the shape of the limbs and nature of the hinge zone, although adjacent layers may show widely varying styles (*see* Figure 6.2A).

The shape of the folded layer is important in determining how the fold has been formed. Folds where the layers maintain a constant thickness through successive levels of the fold are known as **parallel folds** (Figure 6.4C). They are also sometimes referred to as **concentric folds** because the curved surfaces of each successive folded layer, which are parallel to each other, have the same centre of curvature. Such folds are formed by a process known

as **buckling** where a relatively strong layer has been subjected to compression acting approximately parallel to the layer; Figure 6.2B is a good example. The folded layers must be deformed internally (i.e. strained) for bending to take place, as discussed below.

In **similar folds**, by contrast, (Figure 6.4D) the distance across each layer, measured parallel to the axial plane, is constant and each layer has exactly the same profile (e.g. *see* Figure 6.2E). In such folds the layers thicken in the hinges and thin in the limbs in a regular manner.

Chevron folds (Figure, 6.4E) are characterised by straight limbs and sharp, angular hinges. Although the individual layers are generally parallel, the folds overall have a similar geometry (*see* Figure 6.2C).

A whole range of possible fold shapes exists in nature, embodying transitions between these 'end-member' types. In most folds, the layers differ in thickness to a greater or lesser degree and often show a variation between a more parallel shape in the thicker, stronger layers, and a more similar shape in the thinner, weaker layers. As an example, compare the thick layers in the lower fold of Figure 6.2B with the thinner layers in the core of the fold.

Fold sets may be described as **symmetric** if the opposite limbs are of equal length, or **asymmetric** if unequal (Figure 6.4F).

The folding process
Parallel folds
These folds are restricted by their geometry to a depth that is determined by their wavelength, as shown in Figure 6.5A, and must be replaced by a different mechanism at their centre of curvature. This means that perfectly parallel folds (which are actually a set of arcs of concentric circles) are restricted to a given thickness of layers. Consequently, we can assume that it is the thickness of the (strong) layers involved in the folding that determines their wavelength, and by knowing that, an estimate of the original thickness of the layers involved can be deduced. For example, the very large-scale folds seen in Figure 6.1, which have a wavelength of several kilometres, must have affected a particularly strong layer whose thickness could have been up to approximately half this amount. Such folds may rest on a detachment surface, such as a thrust plane, separating them from a different set of rocks beneath the parallel-folded layer, or alternatively,

the shortening may be taken up by a different mode of deformation in the rock beneath, as in Figure 6.2B.

Similar folds
Perfectly similar folds are characteristic of very **ductile** (i.e. weak) material where the individual layers behave passively under compression and accommodate themselves to a new shape determined by the constraints imposed by the geometry of the surrounding stronger material. The process of deformation here is described as 'flow', using the analogy of liquid behaviour, although of course the movement takes place in the solid state. The direction of flow is across the individual layers, and in the ideal case, parallel to the axial

plane of the fold as in Figure 6.5B. The process can be visualised by drawing a set of lines across the edges of a pack of cards, then shuffling the pack into a fold shape. The movements here are obviously parallel to the individual cards, which themselves are parallel to the axial surface of the fold. Since the ductility (i.e. relative strength) of a rock will depend on the ambient temperature, as explained in Chapter 4, it follows that similar folds are much more common in rocks deformed under metamorphic conditions, for example in **shear zones** (*see* below) where the flow direction is parallel to the walls of the shear zone.

The scale on which a fold is viewed is important here, since folds that appear to be similar on a large scale turn out to

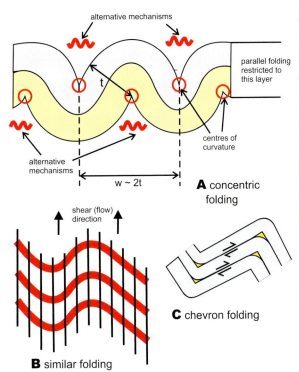

A concentric folding

B similar folding

C chevron folding

Figure 6.5 The folding process. **A.** Parallel or concentric folding: once the folds have reached a semicircular shape, any further tightening will result in distortion of the limbs and the folds will no longer have a parallel geometry; note that below and above the parallel-folded layer, an alternative fold mechanism must operate. The wavelength (w) of the parallel folds at this point equals approximately twice the thickness (t) of the folded layer. **B.** Similar folding: the fold shape is achieved by variable flow across the red layers, which behave passively; the flow direction is parallel to the axial plane of the folds. **C.** Chevron folding: each layer is parallel folded and slides over the layer beneath; gaps are formed at the hinges, which may fill with vein material; however, the overall fold shape is similar.

be made up of a mixture of active and passive layers. Where a set of rock layers with varying strength (e.g. sandstone beds separated by weaker shales) are subjected to compression, the shapes of the resulting folds are determined by the way the strongest layers deform, usually by the process of buckling, whereas the weaker layers tend to accommodate to the shapes formed by the buckled layers. This is shown particularly well in Figure 6.2B, where the fold shape has been dominated by the buckling of the thick sandstone layers, which show nearly parallel geometry, yet the folds as a whole are closer to similar. Even folds on the scale of a hand specimen that look perfectly similar in shape may turn out to contain parallel-folded layers when viewed under the microscope.

Chevron folds

These are a special case of similar folds, and are formed in thinly-layered rocks such as slates, where the strong layers are separated by even thinner weak zones that facilitate the sliding of the layers over one another (Figures 6.2C and 6.5C). The fold limbs are straight and separated by sharp, angular hinges. Note that the folds in Figure 6.2B maintain their shape through the whole width of the outcrop and are thus broadly similar despite the fact that the individual layers are generally parallel. A close look at these folds would reveal that the strong layers have only been able to maintain their shape because of the flow of the weak inter-layer material into the hinge zones (Figure 6.5C).

Strain in folding

As we have just seen, there are a number of different methods by which a fold can be produced. The process of folding involves changes to the geometry of the folded layers, regardless of which particular fold mechanism is responsible; each of these mechanisms involves internal strain in the folded material, but the way the strain is distributed within that material can vary widely. Figure 6.6 shows examples of different types of fold produced by compression parallel to the folded layer, which demonstrate this variation. In folded layers with broadly parallel geometry (examples A–D), the strain distribution depends very much on the physical properties of the folded material. This can be demonstrated by experimenting with easily available artificial materials. Thus folding a stack of paper produces a buckle fold by means of each individual sheet slipping over its neighbour without any visible strain within each sheet. This process is termed **flexural slip** (Figure 6.6A). Using a different material, a thick sheet of rubber or foam plastic, for example, folding is accomplished by concentrating strain at the hinge zone, with extension on the outer arc of the fold and compression on the inner arc, whereas the limbs are unstrained (Figure 6.6B); this process is sometimes referred to as **ideal buckling**. If a slightly stronger material is used, however, such as stiffer rubber, the strain may become concentrated in the fold limbs, leaving the hinge internally almost unstrained (Figure 6.6C); such a process is known as **flexural shear**. In the example shown in the figure, the left-hand limb of the anticline is subjected to dextral shear and the right-hand limb to sinistral shear. In addition to the strain produced by the buckling process, there is usually an element of shortening strain affecting the whole rock, and which increases in importance as the folds become tighter. This shortening strain has the effect of thinning the limbs and elongating the hinges of upright folds, as shown in Figure 6.6D. Ultimately, folds initiated by buckling become almost similar in style after undergoing considerable shortening.

The **chevron fold** style, in contrast, is difficult to replicate under normal conditions but has been produced in the laboratory under high confining pressure. The fold limbs so formed are straight, and rotate in such a way that all the strain is taken up by the weak layers and by the narrow hinge zones, whereas the straight limbs merely slip laterally past each other, without undergoing any internal strain (Figure 6.6E). **Kinking** (Figure 6.6F) relies on the same mechanism as chevron folding but is confined to discrete bands (**kink bands**) that behave in a similar fashion to reverse faults or **shear zones** (*see* below). Based on experimental evidence, it has been suggested that a rock under compression may shorten by means of two sets of kink bands, with opposed inclinations, each pair of which expands laterally, and eventually intersect to form symmetrical chevron folds. However it is doubtful whether all chevron folding is formed in this way.

The way in which a rock layer deforms into a fold shape depends both on the material of which it is made (e.g. how strong it is) and on the surrounding physical environment, especially the temperature and pressure. Deformation at upper levels in the crust takes place at low temperatures and lithostatic pressures, and is dominated by fracturing and faulting. On a microscopic scale,

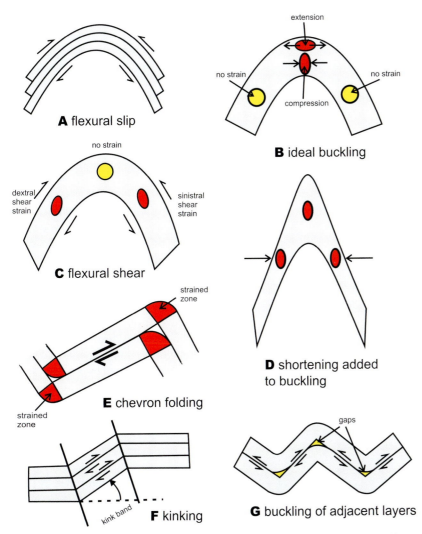

A flexural slip

B ideal buckling

extension

no strain — no strain

compression

C flexural shear

no strain

dextral shear strain — sinistral shear strain

D shortening added to buckling

E chevron folding

strained zone

strained zone

F kinking

kink band

G buckling of adjacent layers

gaps

Figure 6.6 Strain in folding. **A.** Flexural slip: the fold is produced by thin, relatively unstrained layers sliding past each other. **B.** Ideal buckling: strain is confined to the hinge, with extension in the outer arc and compression in the inner. **C.** Flexural shear: the hinge is relatively unstrained but the limbs experience shear strain. **D.** Layer-parallel shortening applied during or after buckling produces a fold approaching similar shape. **E.** In chevron folding, the strain is concentrated in the hinges and the thin, relatively unstrained, layers slide past each other. **F.** Kinking: a kink band is formed by the same mechanism as chevron folding, but the structure is localised and acts like a reverse fault. **G.** Buckling of a set of parallel folded layers can be achieved by leaving gaps at the hinges, which are then filled by vein material.

Figure 6.7 The role of faults in folding. Hinge zone of the Lady anticline, Pembrokeshire, illustrating how small faults (outlined in white) help to accommodate the strain in the hinge zone.

it would be seen that the strain had been achieved by small displacements caused mostly by sliding between the rock particles, rather than by deformation within the grains themselves. These displacements may be accompanied by the solution and deposition of vein material. Where rock layers are extended on the outer curves of parallel folds, as in Figure 6.6B, the extension is achieved by the opening of extensional cracks – look, for example, at the cracks in the thick folded layers in Figure 6.2B; these may then be filled by vein material such as quartz or calcite deposited from solution. Compression on the inner curves of parallel folds causes material to be squeezed out or dissolved, and transferred to the extending areas. Figure 6.6G shows how the buckle folding of adjacent layers may leave gaps that may become filled with vein material. Many folds are associated with faults, movement on which helps to determine the final shape of the fold. The fold shown in Figure 6.7 is a good example

6

FOLDS AND FOLDING

of how the shape changes required by the buckle-folded layers are achieved by displacements along small faults.

In deformation at lower crustal levels, under conditions of elevated temperature and pressure, rocks typically develop a **fabric** (*see* Chapter 7) and tend to be dominated by similar-style folding where the flow direction is across the layers, which merely rotate passively in response. Here the whole rock is involved in the strain due to pervasive recrystallisation; ultimately, each individual grain or particle may have changed shape in order to accommodate to the overall strain.

It is difficult to imagine solid rock behaving in this way, and because the process is so slow, it cannot be observed. Fortunately, however, flow folds of broadly similar style can be seen in glacier ice (Figure 6.8) which does deform in the solid state and over observable timescales. Such flow patterns, therefore, are an exact analogy of the type of folding seen in some metamorphic rocks (e.g. *see* Figure 7.5A).

Multilayer buckle folding

In a set of layers dominated by buckle folding, the wavelength of the individual folded layers may be the same, as in Figure 6.9A, which is the likely outcome if the layers are all of approximately equal thickness and are composed of similar relatively strong material, such as limestone or sandstone. In such a case, the weaker material between the buckled layers has to deform by a different mechanism, such as lateral shear, or flow, controlled by the stronger layers on either side moving in opposite directions. However, a set of folded layers of varying thickness and/or strength will probably respond to buckling in different ways, and the layers may then show a variety of wavelengths, as in Figure 6.9B. If such a set of folded layers is unravelled, the length of the unfolded layers varies, despite the fact that the amount of shortening applied to the set as a whole would have been the same. This, at first sight, paradoxical result is explained by the fact that the thicker or stronger layers experience some (layer-parallel) shortening strain before buckling is initiated, whereas the thinner or weaker layers tend to buckle more readily. The minimum amount of layer-parallel shortening experienced by the set of layers is given by the difference between the longest and shortest unravelled layers as shown in Figure 6.9B.

Figure 6.9C shows the end result when a thin layer, having buckled early in the shortening process, has been subsequently refolded on a longer wavelength controlled by the two neighbouring thicker layers that commenced buckling later. Fold sets displaying markedly different wavelengths are common (e.g. *see* Figure 6.2A) and it will be seen from this

Figure 6.8 Flow folds in glacier ice, Lewis glacier, Alaska. The folds on the surface of the glacier are formed by variously coloured layers of rock debris and other structures such as crevasses, originally oriented transversely, which have been deformed by the solid flow of the glacier ice; the direction of flow is parallel to the walls of the glacier and is greater towards the centre. ©Jim Wark.

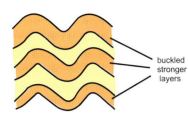

A

buckled stronger layers

C

weaker buckled layer

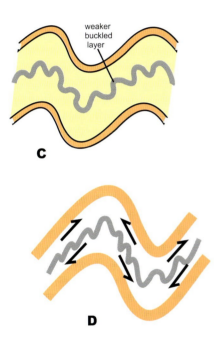

D

X

Y

Z

→ ←

B

minimum shortening in X

minimum shortening in Y

X

Y

Z

hinge this side

unexposed

E

Figure 6.9 Multilayer buckle folding. **A.** In a set of buckled layers of approximately equal thickness and strength, the wavelengths of the layers will be equal, but the weaker material between the layers will deform by flow to take up the space left. **B.** Three layers, X, Y and Z, with differing strength and thickness, experience different degrees of layer-parallel shortening before buckling takes place; the minimum amount of shortening is revealed by unravelling the layers. **C.** A thinner and weaker layer between two stronger buckled layers, will form shorter-wavelength folds that are refolded by the longer-wavelength folds. **D.** Folds as in C after further shortening experience shear strain in the limbs causing the smaller parasitic folds to become asymmetric. **E.** The sense of asymmetry of parasitic folds can be used to determine the (unexposed) hinge position of the major folds; i.e. this must represent the overturned left limb of an antiform, not the right-hand limb of an upright antiform.

example that quite complex fold patterns can be achieved by buckle folding, given some variety in the physical properties of the layers involved.

The small-scale folds that affect the thin layers of a multilayer set such as the central band of Figure 6.9C are often called **parasitic** (or **satellite**) **folds.** As the thicker controlling layers tighten, these parasitic folds will become asymmetric due to the shear strain exerted by the adjacent buckling layers (Figure

6.9D). This feature is useful in determining which direction the main antiformal and synformal hinges are, in situations where the outcrop is limited and the fold limb is steeply inclined or possibly overturned (Figure 6.9E).

Folds in three dimensions

Most of the discussion so far has been based on considering folds in profile, where their geometry is described in a plane perpendicular to the fold

hinge. In reality, of course, folds are three-dimensional objects that may vary considerably in the dimension parallel to their hinge lines, and folds that maintain their profile shape for long distances are uncommon.

Figure 6.1 displays three-dimensional variation in a quite spectacular way. The main anticline in the picture gradually dies out away from the viewer by means of a decrease in height and width, and a parallel anticline to the left of it is shaped like an upturned boat, with an oval ground plan. Between these two structures, and to the right of the main anticline, there are rather oddly-shaped synclinal structures. Folds of this type, which vary in height along their length such as to plunge in opposite directions at each end, are termed **periclines** and may be either anticlinal or synclinal. In the extreme case, an anticlinal pericline where the dips are radial becomes a **dome**, and the synclinal equivalent becomes a **basin**.

Fault-induced folding

Many folds are directly caused by fault movements, as we saw in Chapter 5. Such folds are confined to the upper brittle layer of the crust and are of two main types – **active bending** induced by the upward or downward movement of a layer under compression, and **passive bending** of a layer under extension. Thrust faulting of layered rocks produces hangingwall synforms at the cut-off point at the foot of a ramp and antiforms where the thrust sheet rises over the top of a ramp (Figure 6.10A). More complex shapes result from multiple ramp/flat arrangements or from movements on lower thrust systems (e.g. *see* Figure 5.7). An important feature of the folds produced in this way is that the strain produced

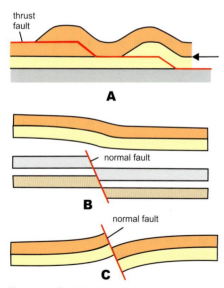

Figure 6.10 Fault-induced folding. **A.** Antiforms and synforms in the hangingwall of a thrust sheet caused by the movement of the sheet. **B.** Drape fold formed above a normal fault. **C.** Hangingwall antiform and footwall synform formed as a result of gravitational accommodation to the fault movement. See Chapter 5 for further explanation.

around the hinge area moves progressively along the thrust sheet as the sheet itself is transported. In consequence, each part of the layer is successively subjected to episodes of both extensional and compressional strain that will inevitably affect the microstructural fabric of the whole layer.

Passive bending folds may form as a consequence of extensional faulting. Two examples are shown in Figure 6.10. An unfaulted layer may form a **drape fold**, which has both antiformal and synformal components, above a normal fault (Figure 6.10B); such folds are gravity driven and are primarily extensional at the antiformal bend and compressional on the synformal bend. Another common class of passive bending structures, discussed in the previous chapter, are the hangingwall antiforms and footwall synforms associated with extensional faulting (Figure 6.10C).

Superimposed folding and interference structure

The core regions of mountain belts typically contain metamorphic complexes, originally formed at deep crustal levels but now exposed at the surface. These may have experienced several episodes of deformation; indeed, structural geologists have been known to record up to nine such episodes in a single metamorphic complex! However, cases of metamorphic terrains having experienced more than two episodes of penetrative ductile deformation that have affected the whole of the terrain would appear to be uncommon. There may have been several episodes of deformation after the last period of thoroughgoing ductile deformation in a region, but these will

generally be localised and have a limited effect on the overall strain pattern.

Where two ductile deformations are superimposed, a series of rather strange shapes can be formed as a result (Figure 6.11). The geometric consequence of such a superimposition is known as an **interference structure**. These structures are typically formed under conditions of elevated temperature resulting in flow folds of similar-type geometry, as in Figure 6.11A. The earlier set of folds in such a structure need not necessarily have formed under these metamorphic conditions, but in order for penetrative ductile behaviour to be imposed during the later deformation, the rocks must either have been retained at deep level or re-buried.

A wide variety of complex shapes are possible with superimposition, depending on the relationship between the fold axes and axial planes of the two systems, and their orientation with respect to the viewer. The examples of Figures 6.11A and B are obvious, but more enigmatic shapes are possible. Two common types are shown as examples: Figure 6.11C shows the outcrop pattern produced by two superimposed upright folds with vertical axial planes, whereas in Figure 6.11D, the F1 folds are tight with an inclined axial plane.

Shear zones

At deep levels in the crust and upper mantle, zones of ductile displacement, termed **shear zones,** take the place of faults at higher levels. These zones of relatively high strain are confined between blocks on each side where the strain is either absent or appreciably lower. They are formed when the boundary blocks or 'walls' of the

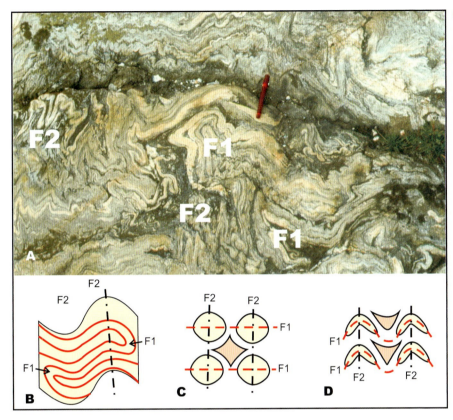

A

Figure 6.11 Superimposed folding. **A.** Complex interference structure formed by superimposition of upright F2 folds on recumbent isoclinal F1 folds in banded gneiss. **B.** Interference structure formed by superimposition of F2 flow folds on isoclinal F1 folds whose axial planes are approximately perpendicular. **C, D**, common patterns found in superimposed flow folds with perpendicular axes; in C, both folds are upright; in D, F1 is a tight overfold (the relationship shown in C gives a pattern of domes and basins often referred to as 'egg-box' pattern).

Figure 6.12 Shear zones. **A.** In a simple shear zone, a zone of ductile strain is formed between two blocks moving in opposite directions; the strain increases towards the centre of the zone. **B.** The displacement caused by faulting in the upper brittle layer of the crust may become transformed into a shear zone in the lower, more ductile crust. Inspired by a diagram by J.G. Ramsay (1987). **C.** Shear zone in previously undeformed granite; a faint fabric (outlined in white) in the granite to the left of the shear zone curves gradually into it; the zone itself is highly strained and much finer-grained. The left-hand side of the shear zone has moved upwards in relation to the right, as in a reverse fault.

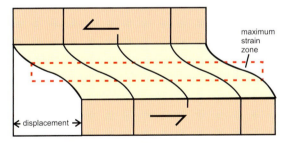

A

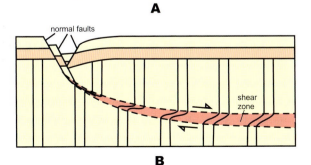

B

C

shear zone

shear zone are displaced in opposite directions in the same way as fault movements (Figure 6.12A). Since they are a relatively high-temperature phenomenon, they are found only in metamorphic rocks, and are particularly important in Precambrian gneiss complexes. It has been suggested that large sub-horizontal shear zones may constitute the typical structure of the lower part of the continental crust.

Shear zones are found on all scales from a few millimetres in width to many kilometres. Some of the largest shear zones of the Precambrian basement of southern Greenland are up to 50 kilometres across and comprise both thrust-sense and strike-slip types. Major faults in the upper brittle crust will pass downwards into shear zones at depth, as shown in Figure 6.12B.

In an ideal shear zone, the strain increases from zero at one wall of the zone to a maximum in the centre of the zone, and decreases to zero again at the other wall. Such ideal zones may be found in otherwise undeformed igneous rocks containing no previous structure, as in Figure 6.12C. In this case, the strain pattern can be seen by the variation in shape and orientation of the deformed crystals. Generally, however, shear zones occur in previously deformed rocks such as gneisses, where the strain can be deduced from the bending and thinning of the gneissose banding.

Why do folds form?

The reasons for the formation of folds may not be immediately obvious. Referring back to the discussion of the behaviour of materials under stress in Chapter 4, the feature that governs whether rock behaves in a brittle or ductile manner was **strain rate** (which in turn is controlled by temperature and pressure); that is, a more rapid strain rate would lead to fracturing, whereas a lower strain rate was necessary to achieve ductile behaviour. Consequently, a necessary precondition for a rock to produce folds would be a low stress (but still above the yield point of the material) applied for a long period of time, thus giving a low strain rate. An applied stress that was too large would merely cause the rock to fracture, which would immediately relieve the stress. It follows that the stress must be applied over periods measured in years rather than seconds, if appreciable ductile strains are to be achieved. Large folds will take thousands of years to form.

Another important precondition for folding is that the rock must have a layered structure – folds cannot form in a homogeneous rock. Ductile strain in an unlayered material would result in microstructural changes spread throughout the whole rock, ultimately producing a **fabric** (*see* Chapter 7). Moreover, the layering has to fulfil certain conditions before folding can take place. Experimental work has shown that buckling will only occur if relatively strong layers are separated by material that is very much weaker (perhaps by a factor of 1/50) capable of flowing in such a way as to accommodate to the shape of the buckled layer (*see* Figure 6.9). Chevron folding, as we have seen, also has quite strict requirements in terms of the physical properties of the folded layers.

Ideal similar folding, on the other hand, requires the rock layering to behave in a wholly passive manner, that is for 'flow' to occur across the layering. If any of the layers are strong enough to buckle, the resulting folds will not be truly similar in style. The conditions for flow folding to occur are that the ductile flow-folded material must be constrained by a layer that behaves actively, that is, the flow folding should be controlled by movement of the boundaries of the flow-folding system. This principle is most easily demonstrated by the example of a shear zone, whose deformation is controlled by the lateral movement of its walls (*see* Figure 6.13A), or the glacier folding of Figure 6.8. The same applies to the flow folding exhibited by weak material constrained between two strong buckled layers – here, again, it is the geometry and movement of the strong layers that control the flow-folding pattern.

These examples illustrate an important principle, which is that folding is largely dependent on the relative movement of blocks of rock; in other words, it is more helpful to understand folding in terms of a **kinematic** (or movement-based) system rather than a **dynamic** (or stress-driven) one. Many fold systems are constrained by major fault surfaces (especially thrust planes – *see* Chapter 10) and their geometry is controlled by the relative movement of the fault sheets that form their boundaries. Going up to a larger scale, it is the convergent plate movements at subduction zones and collisional plate boundaries that drive the deformation in fold belts.

7   Fabric

The changes to a rock brought about by deformation, particularly under metamorphic conditions, may result in the development of a new structural 'texture', termed the **fabric**. At relatively shallow crustal levels, where brittle deformation dominates, fabrics may consist of closely spaced fracture planes, forming a **fracture cleavage**. At deeper crustal levels, deformation is strongly influenced by higher temperatures and the greatly increased lithostatic pressure caused by the weight of rock above. Under such conditions, with the rise in temperature and pressure, the deformation increasingly takes place by means of flow in the solid state; this depends on the reorganisation of the minerals making up the rock layers in such a way as to accommodate to the new shape of the deforming layer. As we saw in the previous chapter, individual layers may deform passively, rather than actively as in the buckling process, and the flow direction may be oblique or even transverse to the layer being folded (Figure 6.5B).

The fabric may consist of changes to the shapes and orientations of grains, crystals and other objects making up the rock body so as to reflect the change in shape (**strain**) of the whole rock body. Thus, for example, particles within the rock may recrystallise so as to become flattened or elongate in the direction of greatest extension, and new minerals may crystallise with their long axes in that direction.

Such changes were illustrated in Figure 4.9, where an undeformed rock body is represented by a cube; the deformed (or strained) body then becomes rectangular-sided, with its longest axis representing the direction of greatest extension and the shortest, the direction of greatest compression. The fabric, therefore, reflects the strain, and shows how the rock has responded to the stresses acting on it. Layers parallel or sub-parallel to the short axis of the deformed body may be folded, and those oriented close to the long axis will be extended. Extended layers may be thinned, or may break up to form structures called **boudins**; here the layer is pulled apart into sections that have the appearance of a string of sausages (this structure, called **boudinage**, is described in Chapter 4 – *see* Figure 4.8).

New planar structures formed in this way are termed **foliation**, and linear structures, **lineation**. Fabrics may be wholly planar (S-fabrics), or wholly linear (L-fabrics), or have both planar and linear elements (SL-fabrics). Planar fabrics include the well-known **slaty cleavage,** together with **schistosity** and **gneissosity**. The latter two structures are found in coarsely crystalline metamorphic rocks, in which the nature of the fabric is obvious, whereas the grains in a slate are so small that their shapes and orientations can only be observed under the microscope.

Cleavage

The term '**cleavage**' covers a variety of different types of planar structure formed in different ways. As the name suggests, a rock will usually cleave or split along the cleavage planes. The more common types are: **fracture cleavage**, **slaty cleavage**, **crenulation cleavage**, **spaced cleavage**, and **pressure-solution cleavage**. These terms are not mutually exclusive, and individual examples may consist of two or more of these types, and may also depend, to some extent, on the scale of observation.

Fracture cleavage

This type of cleavage, as the name indicates, is formed of closely spaced fractures, and is therefore a product of deformation in the upper brittle region of the crust. The fractures may either be extensional in origin, formed at right angles to the extensional stress, or be caused by shear, in which case they have the same geometrical relationship to the stress axes as shear faults (*see* Figure 5.11A); that is, they form at an angle to the maximum compressive stress. A fracture cleavage may act as host to mineral veins, particularly if it is extensional. Since rocks vary widely in strength; it is not unusual to find deformed layered sedimentary rocks in which the stronger, more brittle, layers possess an extensional fracture cleavage, whereas the weaker, more ductile, layers have deformed by viscous flow – see, for example, the boudins in Figure 4.8C, which have been separated initially by extensional fractures. Shear fracture cleavage is often found in fault rocks, in which case it may be a useful indicator of the sense of movement on the fault (*see* below).

Slaty cleavage

This type of cleavage is confined to fine-grained rocks, such as mudstones or volcanic tuffs, which have been deformed under low-grade metamorphism. It consists of closely-spaced planes of weakness that make the rock easily cleaved and suitable for roofing slate (Figure 7.1A). The nature of the cleavage planes is usually only apparent under the microscope, when they are seen to be caused by the parallel orientation of planar minerals such as clays or muscovite mica. There may also be a parallel arrangement of elongate particles or grain aggregates. The origin of the slaty cleavage becomes more obvious when the rock contains objects of known initial shape such as fossils, reduction spots, or pebbles, in which case the cleavage plane is seen to correspond to the plane of flattening of the objects, as in Figure 7.1B. Slaty cleavage will thus have a simple geometric relationship to the maximum compressive stress, and will usually be either parallel to the axial planes of related folds, as in Figure 5.1A, or have a fan-like arrangement, either converging towards, or diverging away from, the core of the fold (Figure 7.2A, B).

To understand why the fan arrangement occurs, it is necessary to return to the section on strain in folding in Chapter 6. Assuming that the slaty cleavage was initially parallel to the XY strain axes, (i.e. it is a flattening fabric) then buckling of a layer with an initially parallel cleavage will result in a convergent fan, as in Figure 7.2A. However, initially parallel slaty cleavage in a layer folded by flexural shear will cause the cleavage to become divergent (Figure 7.2B). Thus, bedded rocks consisting of interlayered **competent** and incompetent layers (sandstones and shales, for example) may display alternating divergent and convergent cleavage fans, as shown in Figure 7.2C.

Figure 7.1 Cleavage. **A.** Slaty cleavage in slate; the cleavage is approximately parallel to the axial plane of the fold. **B.** Slaty cleavage in glacial mudstone enclosing drop-stones. The bedding (parallel to the black line) is inclined to the right and the cleavage (parallel to the red line) is vertical. Note the darker areas (pressure shadows) at the sides of the pale drop-stone showing where lighter material has been dissolved by pressure solution. **C.** Crenulation cleavage in schist. The cleavage does not penetrate the more competent bands at the top of the picture, which are affected by open folds only. **D.** Spaced cleavage in weakly metamorphosed layered siltstone–mudstone. The cleavage is accentuated by darker bands formed by pressure solution. B–D: note coins for scale.

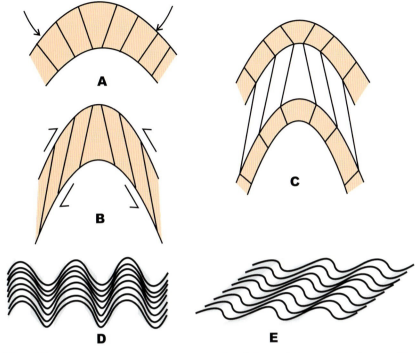

Figure 7.2 A–C, cleavage fans: **A.** Convergent fan: the cleavage planes converge downwards. **B.** Divergent fan: the cleavage planes diverge downwards (parallel to the black lines). **C.** In alternating competent and incompetent layers, cleavage alternates between convergent in the more competent layers and divergent in the more incompetent layer. In competent layers affected by buckling, originally vertical cleavage is rotated, whereas the incompetent layer is affected by flexural shear. **D.** Symmetric and E, asymmetric crenulation cleavage. The cleavage, emphasised by the grey bands, is produced by alternating dark and light bands, the darker bands having been affected by pressure solution.

Effect of pressure solution

All three types of cleavage mentioned above may be modified by the effects of pressure solution and re-deposition of soluble material forming part of the rock body. This effect is discussed in chapter 4 (*see* Figures 4.12E, F). The more easily soluble rock material, such as calcite or quartz, may be dissolved in zones of high compressive stress, and re-deposited in zones of low stress. The removal of light-coloured material in the high-stress zones will result in these zones becoming darker than the surrounding areas, causing the fabric to become accentuated by the formation of alternating darker and lighter bands (Figures 7.1D; 7.2D, E). This effect is particularly common in asymmetric crenulation cleavage, where material from the long limbs of the asymmetric microfolds has been dissolved and re-deposited in the short limbs, as shown in Figure 7.2E.

Spaced cleavage

This purely descriptive term can be applied to any type of cleavage in which the individual fabric planes are obviously separated by zones of uncleaved rock. Figure 7.1D is a good example. Thus all examples of fracture cleavage and crenulation cleavage may also be described as '**spaced cleavage**'. Even slaty cleavage, if seen under the microscope, may appear to be spaced, although the term 'spaced cleavage' is normally applied to cleavages seen at outcrop scale.

Schistosity and gneissosity

Planar fabrics in more coarsely crystalline metamorphic rocks, i.e. those above slate grade, may take the form of schistosity or gneissosity. **Schistosity**

Crenulation cleavage

As the name indicates, **crenulation cleavage** is produced by crenulations, which are closely-spaced microfolds, of the order of millimetres or less in width. Figure 7.2D and E show two examples, formed respectively from symmetric and asymmetric crenulations. The cleavage, when viewed at outcrop scale, has the appearance of a set of bands that cut across the original layering, as shown in Figure 7.1C. The crenulations themselves must be **similar** or near-similar in style (*see* Chapter 5) for the structure to persist through successive layers without any change in shape. Crenulation cleavage is typically accompanied by **pressure solution** (*see* below) which accentuates the fabric by forming alternating darker and lighter bands. Since this type of cleavage requires a set of uniform thin layers for its formation, it is typically found in rocks that have already acquired a strong foliation, such as slates or schists. It is therefore a common product of second or subsequent phases of deformation in multiply deformed metamorphic terrains.

consists of the parallel alignment of planar minerals such as mica and hornblende, which is easily visible to the naked eye (Figures 7.3A, B). It forms various types of schist designated by the mineral that most obviously forms the schistose structure; thus, for example: biotite schist, muscovite schist and hornblende schist. Fine-grained muscovite- or **chlorite** schists that are intermediate in grade between slate and schist are known as **phyllite**. Most, if not all, schists will have started out as slates and become transformed into schists with increasing metamorphic grade. However, the origin of the schistosity as a slaty cleavage, and its relationship to the strain axes, may no longer be obvious unless it corresponds to a plane of flattening, as in Figure 7.3A, or is parallel to the axial planes of related folds, as in Figure 7.3B.

Schists are most commonly formed in **pelitic** or **semipelitic** metasedimentary rocks, that is, those originating as mudstones or siltstones with a relatively high proportion of clay minerals (*see* Appendix). **Psammites** (metamorphosed sandstones), however, may also possess a schistosity, perhaps in thin mica-rich bands, although such a rock, as a whole, would not normally be thought of as a schist.

Gneissosity is a planar fabric consisting of alternating layers or elongate aggregates of light-coloured minerals – mainly quartz and feldspar, and dark-coloured minerals such as mica and hornblende (Figure 7.3C). It often accompanies schistosity in coarse-grained metamorphic rocks that could therefore be referred to as 'schistose gneisses'. Many gneisses are formed from highly deformed, coarse-grained igneous rocks such as granite, in which the quartzo-feldspathic components have been concentrated into bands or lenses as a result of deformation, as in Figure 7.3D. The relationship of the gneissosity to the strain state of the deformed rock is often fairly obvious, particularly in igneous rocks containing phenocrysts or megacrysts that have been deformed into lensoid shapes

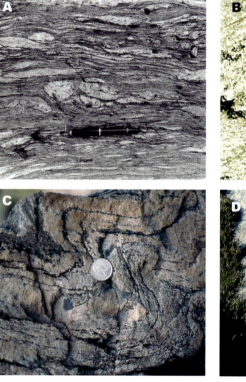

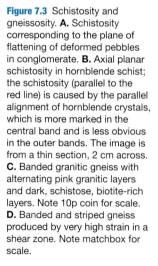

Figure 7.3 Schistosity and gneissosity. **A.** Schistosity corresponding to the plane of flattening of deformed pebbles in conglomerate. **B.** Axial planar schistosity in hornblende schist; the schistosity (parallel to the red line) is caused by the parallel alignment of hornblende crystals, which is more marked in the central band and is less obvious in the outer bands. The image is from a thin section, 2 cm across. **C.** Banded granitic gneiss with alternating pink granitic layers and dark, schistose, biotite-rich layers. Note 10p coin for scale. **D.** Banded and striped gneiss produced by very high strain in a shear zone. Note matchbox for scale.

(termed **augen**), thus mimicking the shape of the strain ellipsoid, and can be used as strain markers (*see* Chapter 4).

Banded gneisses, of broadly granitic composition, are a common component of Precambrian basement complexes and have been the subject of much debate in the past, when many were thought to be 'granitised' metasediments. Banding, on a scale of centimetres, invites comparison with metasedimentary rocks, and close inspection of the nature of the banding, and of the chemical composition of the gneisses, is often necessary to establish their origin. Intense deformation in deep-crustal shear zones (*see* Figure 6.12) can be shown to transform originally homogeneous granites, cut by quartzo-feldspathic veins, into banded gneiss, as in the example shown in Figure 7.3D, and there is no doubt now that most granitic gneiss complexes are in fact igneous, rather than chemically altered metasediments.

Fabric asymmetry and shear sense

Certain fabrics possess an asymmetry that indicates the direction of shear strain (the **shear sense**) of the deformation responsible for creating them; for example, distinguishing sinistral from dextral shear. Such fabrics generally consist of two non-parallel sets of planar elements that are oblique to each other – hence the asymmetry – but may also include objects that have been deformed or rotated in the same sense as the shear strain. Asymmetric fabrics are particularly useful in determining the shear direction of shear zones, and are known as **sense-of-movement criteria**.

Extensional crenulation cleavage ('ecc structure')

This type of fabric consists of a crenulation cleavage that is oblique to the trend of the main foliation, in such a way that tthe smaller angle between the two sets of planes opens towards the direction of shear (Figure 7.4A). The crenulations are asymmetric open microfolds (i.e. with a large fold angle) whose long limbs have been thinned and extended, hence the extensional nature of the cleavage. The sense of shear on the crenulation cleavage is the same as the overall shear sense. Ecc structure is typical of slates, phyllites and schists that have been subjected to relatively low degrees of shear strain in a later deformation, and therefore the axes of the crenulations are usually perpendicular to the shear direction.

S-C structure

This type of structure consists of two sets of planes – the 'S planes', which are the main foliation, and the 'C planes', which are a set of shear planes arranged obliquely to the main foliation (Figure 7.4B, 7.5B). As with ecc structure, the C planes have the same shear sense as the main shear zone, and the acute angle between the two foliations opens towards the shear direction. Unlike ecc structure, however, the C planes may be widely or irregularly spaced and may be either shear faults or narrow shears. S-C structure is common in foliated fault gouge (*see* Figure 5.9A) and may be used to determine the direction of movement on the fault. In schists and gneisses, deformed under higher temperatures, the C planes correspond to narrow shear zones. In that case, both fabrics may result from

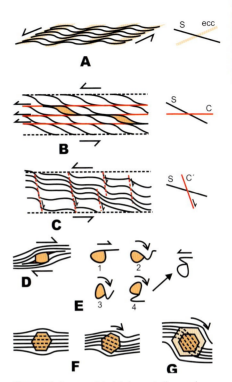

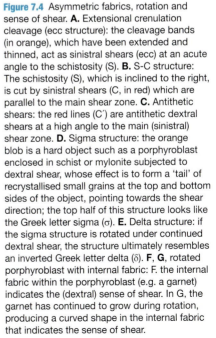

Figure 7.4 Asymmetric fabrics, rotation and sense of shear. **A.** Extensional crenulation cleavage (ecc structure): the cleavage bands (in orange), which have been extended and thinned, act as sinistral shears (ecc) at an acute angle to the schistosity (S). **B.** S-C structure: The schistosity (S), which is inclined to the right, is cut by sinistral shears (C, in red) which are parallel to the main shear zone. **C.** Antithetic shears: the red lines (C´) are antithetic dextral shears at a high angle to the main (sinistral) shear zone. **D.** Sigma structure: the orange blob is a hard object such as a porphyroblast enclosed in schist or mylonite subjected to dextral shear, whose effect is to form a 'tail' of recrystallised small grains at the top and bottom sides of the object, pointing towards the shear direction; the top half of this structure looks like the Greek letter sigma (σ). **E.** Delta structure: if the sigma structure is rotated under continued dextral shear, the structure ultimately resembles an inverted Greek letter delta (δ). **F, G,** rotated porphyroblast with internal fabric: F. the internal fabric within the porphyroblast (e.g. a garnet) indicates the (dextral) sense of shear. In G, the garnet has continued to grow during rotation, producing a curved shape in the internal fabric that indicates the sense of shear.

the same deformation. In progressive shear strain, the XY strain plane (in this case the S foliation) is oblique to the shear direction and gradually rotates towards it (e.g. *see* Figure 4.5E). Minor shears parallel to the shear direction (the C planes) will thus cut and displace the S planes with the same shear sense as that of the main shear zone.

Synthetic and antithetic shears
As we have seen, the C planes in S-C structure display the same sense of shear as the main shear zone, and are known as **synthetic shears**. However, there is often a second set of subsidiary shears that display the opposite sense of shear; these are known as **antithetic shears** (Figure 7.4C). The latter form at a high angle to the main foliation and rotate as the deformation progresses. As Figure 7.4C shows, to enable the rotation to take place, the sense of shear has to be in the opposite sense to the main shear zone. This process can be easily demonstrated by arranging a pile of books such that each book is vertical, as on a bookshelf, then moving the tops of the books to the left or right to simulate a shear zone; as the individual books rotate, the shear sense between them is opposite to that of the shear along their tops. Antithetic shears frequently form along pre-existing discontinuities that cut across the main fabric, such as dykes or veins.

Sigmas, deltas and rotated porphyroblasts
Objects such as **porphyroblasts** contained within the schistosity in a shear zone often exhibit an asymmetry that

Figure 7.5 Asymmetric fabrics and shear sense. **A.** Asymmetric shear folds (see red arrows) indicate dextral shear in highly strained gneiss. **B.** S-C structure in schistose gneiss; the sinistral C shears are inclined to the left at an acute angle to the main schistosity. **C.** Sigma structure formed by sheared magnetite crystals (see red arrow) in mylonite. **D.** Rotated internal fabric (red line) in garnet porphyroblast in mylonitic biotite schist. *Scale:* the lens cap on A and C is ~5 cm across; C and D are from thin sections, each 2 cm across.

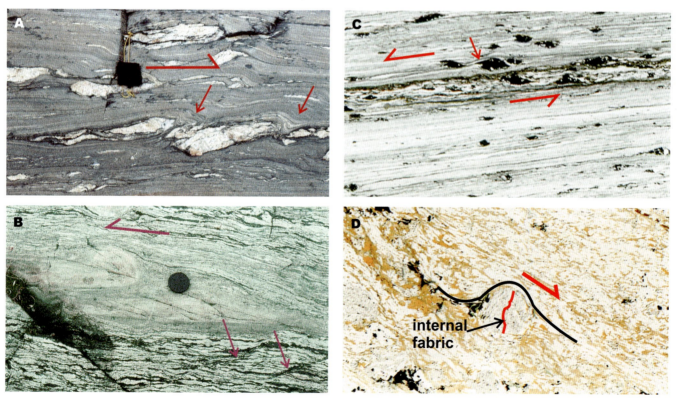

corresponds to the sense of shear of the main fabric. A porphyroblast subjected to shear may grow a 'tail' of small crystals at one or both sides of the object that point towards the shear direction (Figure 7.4D, 7.5C). This is termed **sigma structure** because of its similarity to the Greek letter sigma (σ). If the sheared object rotates in the same sense as the overall shear direction, this gives further evidence of shear sense. However, rotated sigma structure results in the tail of the porphyroblast being bent round in such a way that it eventually appears on the opposite side of the porphyroblast, as shown in Figure 7.4E. This is termed **delta structure**, after the Greek letter delta (δ). Since the tail now appears to indicate the opposite shear sense, care must be taken to distinguish between these two types of structure.

Rotated porphyroblasts often contain an internal fabric that is another source of evidence for the sense of rotation. Garnets are a well-known example of this phenomenon (Figures 7.4F, 7.5D); the sense of rotation is even more obvious when the garnet continues to grow during the shearing process, in which case the internal fabric displays an 'S' or 'Z' shape due to the fact that the last-grown fabric is continuous with the main foliation, whereas the earlier fabric has rotated (Figure 7.4G). It is important to recognise that any perceived rotation is only relative to the enclosing foliation, which itself may have rotated relative to the shear direction.

In interpreting asymmetric fabrics, these should be viewed as nearly as possible in a plane perpendicular to the main foliation and parallel to the inferred shear direction, which will often be visible in the form of an elongation lineation on the foliation surfaces. Viewing fabrics at a high angle to the shear direction will give conflicting or misleading asymmetries.

Lineations

There are several different kinds of lineation, not all of which are structurally significant. A **lineation** is merely a set of parallel lines on the surface of a rock, or which pass through it. Surface lineations, unless they lie on a fabric plane, are difficult to interpret; they usually indicate the intersection of bedding or foliation on a random surface such as a joint, and have no structural significance in themselves. However, surface lineations on two non-parallel surfaces may be used to reconstruct the orientation of the intersecting planar structure, as shown in Figure 7.6A.

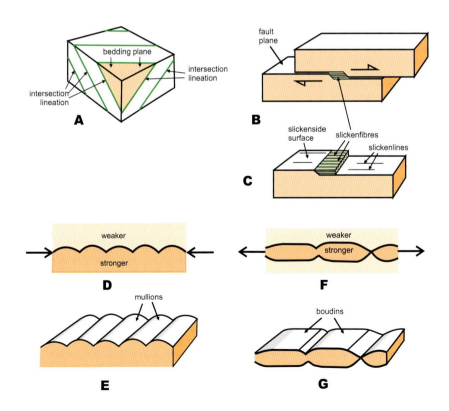

Figure 7.6 Lineations. **A.** Intersection lineations produced on two vertical outcrop surfaces at their intersection with bedding. **B.** Slickenfibres (green) produced by the growth of crystals in a gap between two moving fault walls. **C.** One side of the fault showing the slickenfibres and also slickenlines (grooves) on the slickenside fault surface. **D.** Cuspate structure produced by compression of an interface between stronger and weaker rock layers; and **E**, in three dimensions, this produces mullion structure. **F.** Boudinage (pull-apart structure) caused by the extension of a stronger layer within weaker material; and **G**, in three dimensions, the boudins also form mullion-like structures.

The more 'useful' lineations are linear fabrics that are caused by deformation and are geometrically related to the strain. These may be confined to a single plane or narrow zone such as a fault plane or shear zone (i.e. they are **non-penetrative**), or they may be found throughout a particular piece of rock, in which case, they are said to be **penetrative lineations**. The latter group of structures includes **intersection lineations**, **micro-fold hinges**, **elongation lineations** and **mineral lineations**.

Lineations in faults and narrow shear zones

Movement along fault planes produces two types of linear fabric: **slickenlines**, and **slickenfibres** (*see* Chapter 5, Figure 5.3E). The slickenfibres are elongated crystals, usually of quartz or calcite, that grow from little step-like irregularities on the fault wall and point in the direction of movement of the opposite fault wall (Figure 7.6B, C).

Lineations in narrow shear zones are typically produced by the elongation of objects such as crystals or grain aggregates into near-parallelism with the shear direction, and are therefore equivalent to the elongation lineations discussed below.

Intersection lineations

These are a type of penetrative linear fabric formed by the intersection of two sets of planar fabrics, for example cleavage and bedding (as in Figure 7.7A), or schistosity and crenulation cleavage. Such lineations are usually (but not always!) parallel to fold axes formed in the same deformation event.

Figure 7.7 A. Intersection lineation caused by the intersection of crenulation cleavage with bedding (dipping towards us) in slates. **B.** Cuspate structure formed by buckling of an interface between stronger and weaker layers.

Micro-fold hinges

Folds that are very small in width, such as crenulations, may be considered to form a penetrative linear fabric on the scale of an outcrop.

Elongation lineations

This type of linear fabric consists of a set of objects within the rock that have been deformed in such a way that their longest dimensions are parallel and much greater than their other two dimensions. Objects that were originally near-spherical would then approximate to a prolate ellipsoid, as described in Chapter 4; that is, they represent the X (or maximum) strain axis in a strain ellipsoid such that

$X{>}{>}Y{\approx}Z$. Elongation lineations may be formed from such objects as pebbles, porphyroblasts, grain aggregates, or individual crystals – the latter being known as **mineral lineations**. Elongation lineations are particularly useful in determining the movement direction (shear direction) in ductile shear zones, since at high levels of shear strain, they become close to the shear direction.

In SL fabrics, consisting of both planar and linear elements, the principal strain axes may be uniquely determined. Thus in the case of an elongation lineation lying in a fabric plane of flattening such as a slaty cleavage or a schistosity, the fabric plane is the XY plane, the elongation lineation is parallel to X, and Z is perpendicular to the fabric plane (*see* Figure 4.9).

Rodding and mullion structure

These are sets of essentially cylindrical structures that vary in size from around a centimetre in diameter (**rodding lineation**) to around 10 cm or more across – up to as much as 25 cm. (**mullions**). Rodding lineations are of two main types, the first formed by the extreme elongation of a set of originally more equidimensional objects such as pebbles or porphyroblasts, and the second by the deformation of a thin, hard layer or layers, such as quartz veins, for example, enclosed within softer material. The former type can be distinguished by careful examination, which will show that the individual rods have ends, that is, they are really cigar-shaped, or strongly elongate **prolate ellipsoids** (*see* Figure 4.5B). It is thus a type of elongation lineation as described above. In the case of the latter type, the rods may be formed by extension of the layer; that is, the individual rods are **boudins**, formed by the process of **boudinage**, as described in Chapter 4 (*see* Figure 4.8). Alternatively, the layer may have been deformed by microfolding, in which case the rods may represent crenulation fold hinges, or the intersections of a crenulation cleavage on a foliation surface, as described above.

Mullions are so called because of their resemblance to the architectural features of that name on some older grand buildings such as cathedrals. In their geological context, they are typically formed at the interface between two sedimentary layers of contrasting strength, such as sandstones and shales (or their metamorphosed equivalents). As in the case of rodding lineation, both folding and extension can create a rounded profile at such an interface, which typically assumes a **cuspate** form with the cusps pointing into the more competent material (Figures 7.6D–G). A famous example of mullion structure is recorded in Moine psammites at Oykell Bridge in Sutherland, northern Scotland.

It is obvious from the above that care must be taken to establish the origin of these types of lineation before attempting to interpret their structural significance. For example, an elongation rodding will be parallel to the local maximum strain axis, whereas crenulation hinges may well be perpendicular to it. Of course, both rodding and mullion structure, of whatever origin, may rotate towards the shear direction under high strain, in which case they will become indistinguishable from the elongation lineation.

FABRIC

8   Igneous intrusions

Igneous intrusions occur in a wide range of sizes and shapes. For the present purpose, it is useful to subdivide them into major and minor intrusions, although there is no clear-cut division between them. Minor intrusions include the familiar dykes and sills, which are typically narrow in width, but may extend for many kilometres in length. The major intrusions include very large bodies, especially of granite, whose outcrop extends for tens or hundreds of kilometres in both length and width.

The geological structure of intrusions is important in several ways. As igneous magmas make their way into and through the Earth's crust, their path is partly controlled by the structure of the host rocks, which determines both the shape and the location of the resulting intrusive body. Intrusions may also affect the host rock by imposing an additional stress field on it. Moreover, the process of intrusion may also produce structures, such as planar and linear fabrics, within the body of the intrusion itself, as is now explained.

Structures formed by magma flow

Since magma bodies cool and solidify from the outside inwards, the outer portions are frequently solid or partially solid, while the magma further within the intrusion is still wholly liquid and moving relative to the outer parts. This relative movement may, therefore, create planar or linear fabrics within the intrusion, caused by the parallel orientation of planar or linear elements such as crystals or inclusions. These fabrics will normally be parallel to the walls or roof of the intrusion and are a useful guide to the direction of magma flow. Elongate **phenocrysts** within still-liquid magma are often oriented in this way. The outer parts of an intrusion that are already solid may become fractured and injected by a later magma pulse, which may form intrusive dykes or veins in the older material, or break off and incorporate pieces of it as inclusions. Linear fabrics in dykes are interesting in providing evidence for the propagation mechanism. For example, many long dykes appear to have been propagated laterally away from their source, since the direction of magma flow as indicated by the linear fabric is oblique or even horizontal, rather than vertical.

Classification of igneous bodies

The structural classification of igneous bodies is based on their size, shape and structural relationship to their host rock.

Minor intrusions

The main types of minor intrusion are sheet-like bodies with a long and narrow outcrop, comprising the **dykes** and **sills**, and tube- or pipe-shaped bodies, usually known as **plugs**. **Dykes** are distinguished by their discordant (cross-cutting) relationship to their host rock (Figure 8.1A). They are also, typically, steeply inclined, unless subsequently affected by folding. Steeply inclined sheet-like bodies cutting structurally homogeneous rock such as granite would also be termed dykes.

Sills, by contrast, are generally concordant with the host rock structure (Figure 8.1A); for example, they are parallel to bedding in sedimentary rocks, although in detail they may exhibit short discordant sections, as in the sill shown in Figure 8.2C. Sills in undeformed or weakly-deformed bedded rocks are thus horizontal or gently tilted, and gently-dipping igneous sheets in structurally homogeneous rocks would also be regarded as sills. To summarise, therefore, dykes in their undeformed state are discordant and/or steep, whereas sills are concordant and/or gently inclined.

Most dykes and sills are in the range 1–100 m in thickness, although both thinner offshoots and much thicker examples also occur. One of the largest known dykes – the Great Dyke of Zimbabwe – is up to 11 km thick and extends for around 600 km.

Plugs are steeply inclined or vertical discordant bodies (Figure 8.1B), typically with a roughly cylindrical shape; they are formed in volcanic complexes as the feeder bodies of volcanoes and often occur within **volcanic vents**. Plugs vary from a few tens of metres to around 1 km in diameter.

Major intrusions

Igneous bodies that are large in all three dimensions are referred to as **plutons** (named after the Greek god of

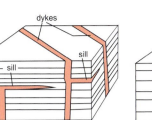
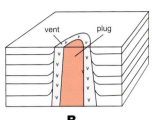

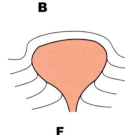

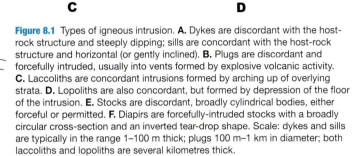

Figure 8.1 Types of igneous intrusion. **A.** Dykes are discordant with the host-rock structure and steeply dipping; sills are concordant with the host-rock structure and horizontal (or gently inclined). **B.** Plugs are discordant and forcefully intruded, usually into vents formed by explosive volcanic activity. **C.** Laccoliths are concordant intrusions formed by arching up of overlying strata. **D.** Lopoliths are also concordant, but formed by depression of the floor of the intrusion. **E.** Stocks are discordant, broadly cylindrical bodies, either forceful or permitted. **F.** Diapirs are forcefully-intruded stocks with a broadly circular cross-section and an inverted tear-drop shape. Scale: dykes and sills are typically in the range 1–100 m thick; plugs 100 m–1 km in diameter; both laccoliths and lopoliths are several kilometres thick.

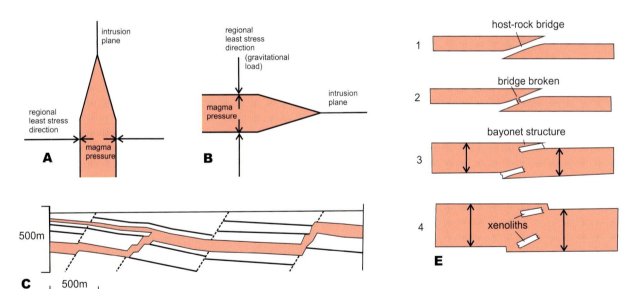

Figure 8.2 Methods of emplacement. Dilational emplacement of a dyke (**A**) and (**B**) a sill; intrusion can occur only if the magma pressure exceeds the host-rock stress acting on the intrusion wall; the intrusion plane will therefore normally be perpendicular to the regional least stress direction, which, in the case of a sill, will correspond to gravitational pressure. **C.** A sill that is concordant with dipping strata may step up along faults in order to maintain the same crustal level (based on a cross-section through part of the Stirling Castle sill, central Scotland, based on McGregor & McGregor, 1948). **D.** The extension direction is given by matching irregularities on each side of a dyke or sill. **E.** Bayonet structure (E3) is formed by breaking of a bridge of host rock between the ends of two overlapping dykes (E1–2), allowing the dyke to expand; pieces of the bridge may subsequently break off to form xenoliths (E4). Based on Nicholson & Pollard (1985).

the underworld), and are also divided into broadly discordant and broadly concordant types. Broadly concordant plutons include laccoliths and lopoliths; these are both roughly lensoid or disc-shaped, the former being convex upwards and the latter convex downwards (Figure 8.1C, D). Laccoliths are usually around 5–10 km in diameter, whereas lopoliths are considerably bigger – up to several hundred kilometres across. There probably exist examples of bodies intermediate between these two types, or even that vary laterally between them, so that the distinction may be rather artificial. Smaller laccoliths may be indistinguishable from sills.

The largest bodies, known as batholiths, have an areal extent of many tens to hundreds of kilometres. Once thought to continue down to great depths in the crust, many have now been shown to have relatively shallow roots, although still perhaps several kilometres in depth. Stocks are a smaller type of pluton, with a broadly cylindrical shape, and discordant relationships to the host rocks (Figure 8.1E). Stocks are typically formed at higher levels of the crust, whereas those deeper-level plutons of comparable size that have been examined in detail have often been shown to have a shape rather like an inverted tear drop, as shown in Figure 8.1F; such bodies are known as diapirs.

Methods of emplacement

As igneous bodies make their way through the crust, space has to be made to accommodate them, and this may be accomplished either by the host rocks moving aside passively, which is termed permitted emplacement, or

by the host rocks being actively forced aside, known as forceful emplacement. To distinguish between these alternative methods requires an examination of the structures of the host rock around the margins of the intrusion and of the internal structure of the intrusions themselves.

Dykes and sills

Most dykes and sills are emplaced by a process of dilation (expansion) of the host rocks; that is, the walls of the intrusion move aside to make way for the magma (*see* Figures 8.2A, B; 8.3B). This process occurs when the regional stress field is such that one of the principal stress axes, at right angles (or at a large angle) to the intrusion wall, is either extensional, or is less than the lateral pressure exerted by the intruding magma, as shown in Figure 8.2A. Sills tend to form at higher levels of the crust where the least principal stress (gravity) is vertical (Figure 8.2B), and dykes rising through the crust may feed sills, as shown in Figure 8.1A, when they reach a level where the magma pressure exceeds the gravitational pressure. Sills are typically concordant with bedding in sedimentary rocks but may also follow faults, as in the example shown in Figure 8.2C. Here, the sill maintains a roughly constant depth by alternately following the inclined bedding and stepping up along the faults.

Careful examination of the walls of a dyke or sill can reveal small steps or irregularities that can be matched on the opposite wall, as shown in Figures 8.2D and 8.3C, from which the direction of opening of the fissure occupied by the intrusion can be deduced.

Some dykes exhibit small vein-like

offshoots, called bayonet structures (Figure 8.3C, D), that reveal how the process of intrusion proceeded. Figure 8.2E shows a sequence of events to explain the observed structure. Two small magma veins whose ends overlap are separated by a bridge of host rock that is subsequently broken through by the pressure of the magma, leaving the characteristic bayonet structure at the margins of the now broadened dyke; pieces of the broken bridge may subsequently float into the magma to be found as inclusions.

Plugs

These types of intrusion are forcefully emplaced, usually through fractured or brecciated volcanic rocks, and are typically preceded or accompanied by explosive activity that has blasted a path through the overlying rock.

Laccoliths and lopoliths

These lens-shaped intrusive bodies are similar geometrically to sills, only on a much larger scale, and their emplacement is achieved in the same way, by dilation of the (usually bedded) host rocks. Laccoliths are considered to have formed by the arching up of the overlying rocks, as shown in Figure 8.1C, in which case the gravitational load must have been exceeded by the magma pressure. Lopoliths, on the other hand, are thought to have formed by the depression of the floor of the intrusion (Figure 8.1D) caused by the weight of the igneous material. Both laccoliths and lopoliths are typically formed from basic magmas. Certain well-known lopoliths are extremely large bodies; the famous Bushveld complex of South Africa is about 300 km in diameter and about 15 km thick, much

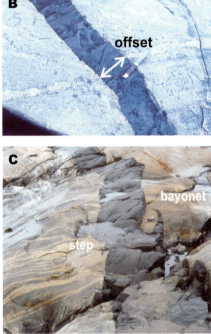

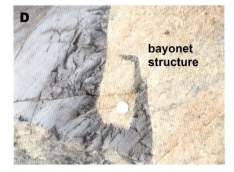

Figure 8.3 A. Dykes are often less easily eroded than their host rocks and stand out like walls, as in this example from Ardnamurchan, western Scotland. **B.** Offset markers, such as this granite vein, indicate the direction of extension of the host rock to allow emplacement of the dyke. **C.** The margins of this dyke show both a step and a bayonet structure. **D.** Close-up view of the dyke in C showing the bayonet structure (see Figure 8.2E).

Batholiths, stocks and diapirs

How large **batholiths**, typically formed of granite, are emplaced into the crust (the so-called '**space problem**') has been the subject of considerable speculation and debate in the past. Some of these bodies are extremely large – the Coast Range batholith of British Columbia, Canada, is around 1,000 km long and up to 150 km across. Batholiths usually appear to be steep-sided, with generally discordant relationships to the host rocks, and are a typical feature of continental margins above subduction zones. Formerly thought to extend to great depth in the crust, some have now been shown to have a relatively shallow base, perhaps only around 10 km deep; however, because they are deeply eroded, their original thickness may have been much greater.

From their relationships to their host rocks, which are typically deformed by them, it is clear that batholiths are to some extent forcefully intruded. However, it is equally clear that forceful emplacement plays only a partial role in the emplacement mechanism, and that space must have been made for these bodies in some other way. In the 1950s and '60s it was thought by many geologists that metamorphosed host rocks of suitable composition could be converted into granite (a process known as 'granitisation') and that this could solve the space problem. However, this process subsequently became discredited, and most granitic bodies are now believed to be genuinely igneous in origin, although minor transformation of host rocks may have taken place around their margins.

It is considered that batholiths are most likely to have been emplaced by a process similar to that responsible for laccolith intrusion, though on a much larger scale, space being found by the relative displacement of either the sides or the roofs of the bodies. Thus smaller, perhaps dyke-like, or pipe-like, bodies of magma may intrude upwards to feed larger bodies at mid- to upper crustal levels, where the host rocks can more easily be displaced upwards or sideways to make the necessary space. Some granite bodies are known to have invaded major strike-slip fault zones or shear zones, where gaps can be easily created at extensional bends, as explained in Chapter 5 (*see* Figure 5.8). Those batholiths that have been investigated in detail have invariably been shown to be composed of many separate plutons emplaced over a long period of time. The mode of emplacement of such bodies, which will now be discussed, is thus key to understanding the formation of batholiths.

of which consists of relatively dense basic and ultrabasic rock, which makes the suggested mechanism plausible.

Stocks and diapirs (Figure 8.1E, F) are much smaller bodies than batholiths, and are often given the generic name of **plutons**. They typically have a broadly circular or elliptical outcrop, 10–20 km in diameter. Some high-level plutons can be shown to have arisen from a larger body such as a batholith at depth; these are usually referred to as **stocks**. The line of granite stocks extending across SW England from Dartmoor to the Scilly Isles is a well-known example of a series of stocks fed by a batholith, which is deduced from gravity measurements to lie at depth beneath them. Other plutons, at deeper crustal levels, appear to have combined to create batholiths.

Many plutons that have been studied in detail are composed of several separate intrusions, each slightly different in composition, which form a series of concentric rings or partial rings, becoming younger inwards. This arrangement suggests that the pluton has gradually become larger over time by successive injections of magma as it moves upwards through the crust, like the inflation of a balloon. Such a process is known as **diapirism** and the resulting pluton is termed a **diapir** (Figure 8.1F). Not all diapirs need have been entirely liquid when emplaced; because of the relatively low density of granite, the effect of the smaller gravitational load compared to that on the surrounding rocks may be sufficient to cause the granite body to rise to a higher level in the crust. This process requires the host rocks to be squeezed aside in order for the pluton to proceed upwards, so that diapirs must be regarded, to some extent, as forceful

intrusions, and the host rocks should show the results of the resulting deformation around their margins.

Stock-like plutons that do not display signs of marginal deformation are more likely to have resulted from permitted, rather than forceful, emplacement. Some stocks, such as the well-studied Devonian granite plutons of the Central Highlands of Scotland, appear to be of this type. The famous Glencoe '**cauldron subsidence**' is an instructive example of how this process may work. The Glencoe structure (Figure 8.4A) is a circular fault surrounding a down-faulted area of Devonian lavas and underlying basement rocks, interpreted as a volcanic caldera. The central depressed area is surrounded by a partial ring of granite that has risen up along an outwardly dipping ring fault, as shown in Figure 8.4B. In this example, the space for the granite has been created by the depression of the cylindrical block, which has sunk into the granite beneath. The granite of the surrounding ring intrusion would have been responsible for much of the vulcanicity within the cauldron.

Several granite stocks in the same general area, including the Cruachan–Starav pluton to the immediate southwest (Figure 8.4A), are composed of several ring intrusions surrounding a central circular body. The structures in the host rocks continue right up to the margins of these bodies without any sign of deviation, suggesting that they are permitted intrusions. They are thought to have resulted from a similar process to the Glencoe structure but at a deeper level, as shown in Figure 8.4C, where a cylindrical block has sunk

beneath the present exposure level, allowing successive batches of granitic magma to fill the space created above the subsiding block. This process is a large-scale example of **stoping**, where blocks of host rock are broken off the roof of an intrusion and sink into the magma (Figure 8.4D). Many granites contain abundant inclusions of host rock, especially around their margins, indicating that this process may have played an important role in their emplacement. The main outer granite of the complex, the Cruachan granite, has also invaded the Glencoe cauldron, indicating that the latter has sunk into the Cruachan magma chamber. The margin of this granite is much more irregular than that of the inner granites and shows extensive veining, melting and assimilation of host rock.

A good example of a more deep-seated pluton, interpreted as a **diapir**, is exposed a short distance west of the Cruachan–Starav complex, on the north-west side of the Great Glen Fault (Figure 8.5A). This body, the Strontian pluton, is a forceful intrusion composed of three main granodiorite components, and has deformed the Moine metasediments around its margins. The flow foliation within the two outer granodiorites defines a synclinal shape, which can be interpreted as the lower half of a diapir, as shown in Figure 8.5B; this shape has then been distorted by further inflation caused by the intrusion of the inner granodiorite. The intrusion is located at the intersection of the Loch Quoich lineament and the Great Glen fault, which is thought to have provided an extensional gap through which the granite magma could flow upwards

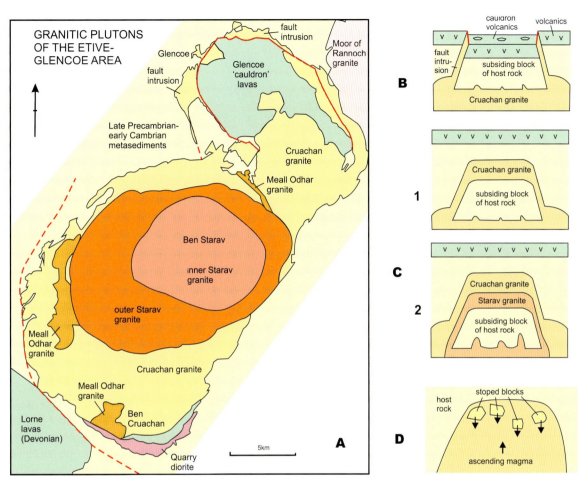

Figure 8.4 Granitic plutons of the Etive–Glencoe area. **A.** The late-Caledonian Starav–Cruachan granite complex consists of five successive intrusions, becoming younger inwards: the Quarry diorite, only present in the SE corner of the complex; the Meall Odhar granite, present as three inclusions within the later Cruachan granite, which forms an almost complete ring around the two youngest granites, the outer and inner Starav granites, both with roughly circular outcrops. To the north is the Glencoe 'cauldron subsidence' where a Devonian volcanic sequence resting on a basement of late Precambrian–Cambrian metasediments has been down-faulted along two 'ring faults'; these have been invaded by granitic 'fault intrusions' and by an extension of the Cruachan granite. **B.** The latter are upper-crustal, largely permitted, intrusions emplaced as ring-shaped bodies in the space created by the down-faulting of a central block. **C.** Two stages of the suggested mode of emplacement of the Cruachan–Starav complex: a central block of host rock has subsided along a sub-surface ring fault, allowing the granites to fill the space created above the subsiding block. **D.** Stoping is the process whereby blocks of host rock are broken off the roof of a pluton to enable the magma to proceed upwards. 8.4A, B based on Johnson (1966).

to form the diapir. Flow foliation patterns in other granite plutons have a convex dome shape that may represent the upper half of a diapiric structure.

Central igneous complexes: ring dykes, cone sheets and radial dykes
Some high-level stocks are distinguished by having an array of minor dyke-like intrusions around them that appear to be genetically related to the stocks.

Such groups of intrusions are known as **central igneous complexes** and are interpreted as the roots of volcanoes. Good examples of this kind of structure are represented in the Cenozoic volcanic centres of western Scotland,

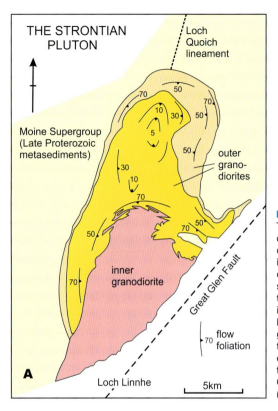

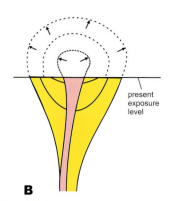

Figure 8.5 The Strontian pluton. A. This late Caledonian pluton consists of three main intrusive phases, two outer hornblende granodiorites and an inner biotite granodiorite. The attitude of the flow foliation in the outer bodies suggests that the outcrop represents the lower half of a diapir, as indicated in **B**, which has become further inflated by the injection of the later inner granodiorite. It has been suggested that the intrusion was focussed on an extensional gap at the intersection of the Loch Quoich lineament and the Great Glen fault. Based on Hutton (1988).

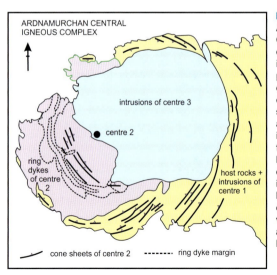

Figure 8.6 Cone sheets of the Ardnamurchan Central Igneous Complex. The Ardnamurchan complex in western Scotland is composed of three separate plutonic igneous centres, of which centre 2 displays an excellent example of concentric cone sheets, each of which occupies an arc of a circle and is inclined inwards towards the centre of the complex at a depth of several kilometres. These cut a set of ring dykes, shown in purple. The later intrusions of centre 3, shown in blue, obscure the eastern half of centre 2. The host rock, together with the earlier centre 1 intrusions, are in yellow. Based on Emeleus & Bell (2003).

such as Skye, Mull and Ardnamurchan (Figure 8.6). In some of these volcanic complexes, a central stock, often composed of multiple ring-shaped intrusions, or **ring dykes**, is surrounded by a concentric set of inclined dykes, called **cone sheets**, or by a set of **radial dykes**.

The cone sheets, which have an arc-like outcrop, lie on a set of conical surfaces that converge approximately at a point source beneath the centre of the igneous complex. It is believed that cone sheets are injected into conical shear fractures formed by the pressure of a suddenly expanding magma chamber, as shown in Figure 8.7A. Selection of steep, outwardly inclined conical fractures by the magma will give rise to ring dykes. Expansion of the magma chamber can also explain the set of radial extensional fractures that come to be occupied by the radial dykes, which also converge at the centre of the complex (Figure 8.7B).

As radial dykes are traced away from an igneous centre, and come under the influence of the regional stress field, they become more parallel and form a **dyke swarm**, which normally forms at right angles to the direction of regional minimum principal stress (Figure 8.7C). Some of the dykes belonging to the Cenozoic swarms of western Scotland extend for long distances from the igneous centres; members of the Mull swarm can be traced for around 200 km, reaching into NE England. One of the largest known dyke swarms is the **Mid-Proterozoic** (ca. 1270 Ma old) Mackenzie swarm of northern Canada, which can be traced south-eastwards from the Arctic Ocean for a distance of over 1500 km. A walk along the shore in parts of western Scotland such as the islands of Arran, Skye and Mull will encounter numerous dykes belonging to these Cenozoic swarms, varying from perhaps one to several metres in width individually, but collectively representing a significant degree of crustal extension.

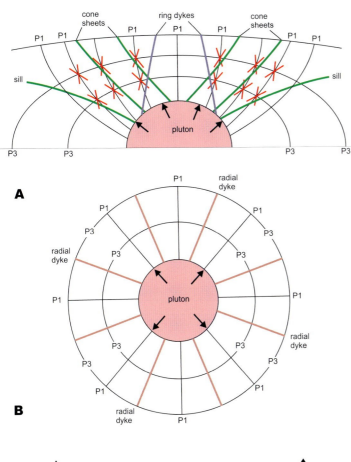

A

B

C

Figure 8.7 Mode of formation of cone sheets, ring dykes and radial dykes. **A.** Vertical cross-section through an idealised central igneous complex. The magma pressure acting at right angles to the pluton margin produces a radial arrangement of the lines of maximum principal stress, P1; the lines of minimum principal stress, P3, which in this case is likely to be an extension, are perpendicular to P1 and follow an elliptical path. As explained in Chapter 5, shear fracture planes (shown here in red) make an acute angle with P1 (see Figure 5.9). It is thought that the most likely method of generating cone sheets is that a sudden increase in magma pressure within the pluton causes shear fractures to form, which, because they make a high angle with P3, become filled with magma. Some sheets may follow a shallower path to form sills. Ring dykes, shown here in purple, may form along steeper fracture planes inclined steeply outwards. **B.** Plan view showing how the lines of maximum principal stress, P1, radiate outwards from the pluton; the minimum principal stress, P3, follows a set of circular paths around the pluton. If P3 is extensional, as is likely if the pluton is expanding, this may create the conditions for radial dykes to be intruded. **C.** Dyke swarms are sets of parallel dykes that originate from an igneous centre; when away from the local stress field generated by the pluton, they come under the influence of the regional stress field and become more parallel.

Structural effects of gravity

The force of gravity affects all geological structures to a greater or lesser degree, and as explained in Chapter 4, is a component of all natural stress fields. However, there are certain types of structure where gravitational pressure is the controlling influence, and it is these that are the subject of this chapter.

Gravitational effects caused by steep slopes

All steep slopes are potentially unstable; we are all probably familiar with examples of cliff collapse and landslides where the rock has been weakened, perhaps by heavy rain, and slid downhill under the influence of gravity alone. Steep slopes are particularly vulnerable to collapse when subjected to earthquake vibration. Movement of material down-slope varies from the

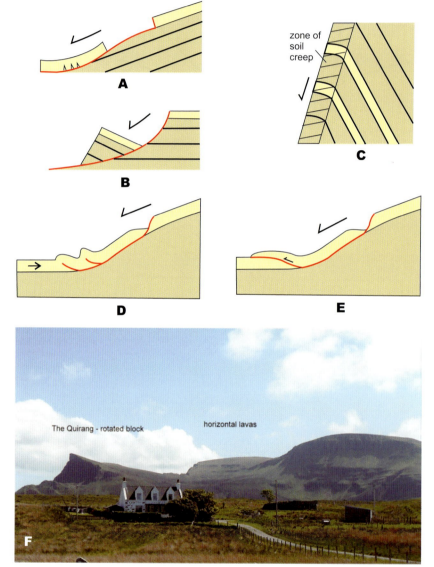

Figure 9.1 Gravitational instability of steep slopes.
A. A strong, cohesive layer overlying weaker material is undercut by erosion and slides down a steep slope. **B.** The outer part of a cliff breaks off and rotates as it slides down a listric fault. **C.** An outer layer of bedrock on a steep slope is subjected to soil creep, where the weaker material flows down-slope causing the stronger beds to rotate downhill. **D.** Gravity-driven down-slope movement of a strong layer is resisted at the foot of the slope, causing deformation at the end of the moving block. **E.** Downward flow of the moving block as in D is aided by a thrust cutting up to the surface. Red lines are faults. **F.** View from the north showing the Quirang back-tilted block on the left with the horizontal Cenozoic lavas on the right; Trotternish peninsula, NE Skye.

simple flow of individual rock fragments to form a **scree** pile or **debris apron**, to quite complex structures that involve folds and/or faults in the rock layers. Some examples of the latter are shown in Figure 9.1. For such structures to form, a plane or surface of easy slip is required to enable the deformation to take place; this slip surface may already exist, in the form of a weak bedding plane, for example, or it may be created in the form of a fault. Some of these structures are indistinguishable geometrically from those formed under an extensional stress regime, as shown in Figure 5.4A and B, the only difference being that the former are formed at the surface.

Soil creep is a process common to many steep slopes and occurs due to the flow of small rock or soil particles down-slope under gravity. This causes any discernible planar surface, such as bedding, to rotate down-slope (Figure 9.1C) and can seriously mislead an observer if taken as the true inclination of the local strata.

Spectacular examples of large landslip structures are to be seen along the north-east side of the Trotternish peninsula in northern Skye, NW Scotland, where huge tilted blocks of Cenozoic lavas have slid down-slope on a layer of weak Jurassic beds (Figure 9.1F). One of these blocks forms the famous 'Old Man of Storr' which has been eroded into a 50 m-tall pinnacle. Blocks of this kind that slide onto a free ground surface will move very quickly, in the order of seconds usually, but if the movement is resisted at the lower end, it may progress at a much slower rate, more akin to the creep behaviour described in Chapter 4.

Large submarine slope failures

The structures described above are on a scale of metres to tens of metres and are of only local significance. However, much larger submarine slope failures have been recorded along both active and passive continental margins. Large slope failures of Pleistocene (late Cenozoic) age recently investigated in southern Chile involve volumes of displaced material amounting to several hundred cubic kilometres. Such active-margin structures are thought to be activated by earthquakes related to the subduction process.

Very large slope failures have also been recorded along passive margins, a famous example being the **Storegga slide** off the coast of Norway, with an estimated displaced volume of over 3000 km^3, along ~ 290 km of coastal shelf. The volume of deposited material has been compared to an area the size of Iceland covered to a depth of 34 m and is attributed to a build-up, on the edge of the continental shelf, of fluvioglacial sediment from the last Ice Age. This collapse is believed to have triggered a huge tsunami in the North Atlantic, which has left traces around the coast of Scotland in the form of a layer of debris, dated at 6100 BCE, in raised beach deposits ~ 4 m above current sea level, and must have had a catastrophic effect on the contemporary Mesolithic population.

Down-slope movement of large fault-bounded sheets under gravity

Large sheets of rock (**nappes**), many kilometres in extent, that rest on a basal fault, may experience sliding under gravity (**gravity sliding** or **gravity gliding**). Sliding can take place on slopes of only a few degrees provided that the weight of the fault sheet is supported by a high enough fluid pressure within the fault zone (*see* discussion in Chapter 5). The down-slope progress of the nappe will be resisted by the rock mass at the end of the slope, i.e. beyond the toe of the nappe, unless the basal fault escapes onto the ground surface (compare Figures 9.1D and E) in which case the nappe may move as a coherent block, as in Figure 9.1E. More probably, it will become internally deformed by folding and/or faulting because of the pressure exerted at its lower end, as in Figure 9.1D. Where the sliding nappe is unrestricted by such resistance, it may travel at high speed, comparable with that experienced in catastrophic landslides, but in the more usual case of restricted movement, the rate of deformation will correspond to that of 'normal' ductile flow in the solid state.

Examples of gravity gliding nappes have been described from various mountain belts, especially the Western European Alps, which contain some of the best-known examples, one of which is illustrated in Figure 9.2. The initiation of the gliding process in these situations is usually a response to crustal thickening due to compression in the central part of the mountain belt. Such a process, involving perhaps many thrust sheets piled on top of one another, is gravitationally unstable and will eventually lead to a reduction of crustal thickness by the lateral spreading of the over-thickened crust; this is termed **gravity spreading**.

Whereas gravity gliding relies on a concentration of movement on a basal fault, gravity spreading implies

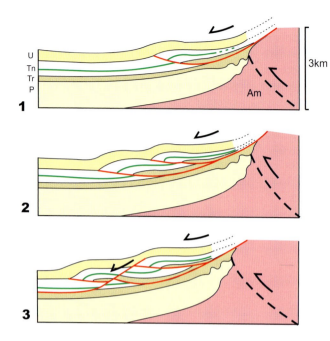

Figure 9.2 Gravity-gliding of Alpine nappes. Three stages in the evolution of the Mesozoic Tinée nappes in the Alpes Maritimes of SE France, from an interpretation by R.H. Graham (1981). **1, 2:** upthrust of the Argentera massif (Palaeozoic basement) causes gravitational sliding of Triassic to Cretaceous strata on detachment faults within a ductile Triassic salt layer in two stages, followed by (**3**) displacement on a listric normal fault. Am, Argentera massif; P, pre-Triassic strata; Tr, Triassic; Tn, Tithonian (Uppermost Jurasssic); U, Upper Cretaceous. Vertical relief ~3 km; horizontal length of section, ~7 km.

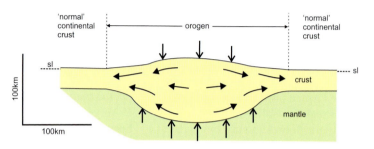

Figure 9.3 Gravitational spreading. Diagrammatic representation of the concept of gravity spreading on the scale of an orogen, or mountain belt: the thickened crust (which can be as much as 80 km compared to an average continental crust thickness of ~33 km) is gravitationally unstable; the resulting stresses squeeze the over-thickened crust, which 'flows' laterally outwards.

largely ductile deformation throughout the moving sheet in response to a gravitationally derived stress state. On the scale of the whole mountain belt, the latter process has also been described as **orogen collapse** (Figure 9.3). On this scale, the over-thickened crust can be viewed as if it is a ductile medium responding to the gravitational stress, which acts to squeeze the crust, by 'flowing' sideways in order to restore the more stable gravitational state of 'normal' crustal thickness. In detail, of course, the resulting deformation is highly variable: brittle, fault-controlled in the upper regions and more ductile creep in the lower.

Glacial flow structure

One of the best examples of gravitational control can be seen in the movement of glaciers. Ice, moving down-slope under gravity, deforms in the solid state, but at a rate that is more convenient for us to study than that of most other rock materials. Structures produced during glacial flow include both folds and faults. A spectacular example of flow folds is displayed in Figure 6.8, where marker bands of differently coloured sediment that have been washed onto the glacier surface have been deformed into a series of similar folds. Fault structures in the form of extensional crevasses also form on glaciers where they move over convex slopes, and these are also deformed by

differential flow into similar folds, as seen towards the top right of Figure 6.8. In both these cases, the flow direction is parallel to the direction of transport of the glacier, but transverse to the structures that have been deformed.

Salt tectonics

Salt, like ice, has a low viscosity compared to most rocks and is therefore able to flow under relatively small stress differences and at geologically fast rates. Common salt (**halite**, or sodium chloride) and **anhydrite** (magnesium sulphate) are the commonest members of a group of minerals, termed **evaporites**, which form by evaporation in warm shallow seas. Evaporite layers are particularly important in the marine Permian and Triassic sequences in Europe. They form convenient glide planes for thrust sheets in the Alps, and in northern Germany a Permian salt layer has developed into numerous salt domes and diapirs. The way these structures develop is illustrated diagrammatically in Figure 9.4. The low-density salt layer is overlain by higher-density sedimentary layers that exert a downward gravitational pressure on the salt. A slight local thickening of the salt layer can focus the salt into a dome-like or anticlinal structure, since the salt can be squeezed sideways by the gravitational load on either side. Nearly a kilometre of sediments is required to produce a sufficiently large gravitational pressure to start the process, but having started, the salt will tend to proceed upwards due to its buoyancy, aided perhaps by extensional fracturing of overlying layers. Once the salt has breached the overlying layer, it forms a **salt diapir** (Figure 9.4.3). When it reaches a level where it is more gravitationally stable, it will spread sideways to form a mushroom-like structure, as shown in Figure 9.4.4, which may eventually become completely detached from its roots.

Salt bodies form a variety of shapes (Figure 9.5A): if the initial disturbance is roughly equi-dimensional, they will evolve into a **salt dome** and then a **salt plug**, which is approximately

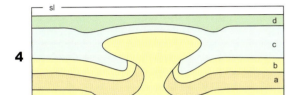

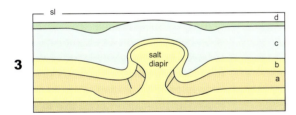

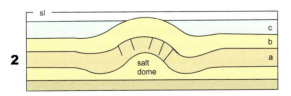

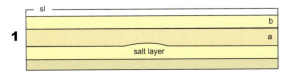

Figure 9.4 Evolution of a salt diapir. The salt layer is gravitationally unstable since it is less dense and has a lower viscosity than the overlying rock. Stage (1): a slight thickening of the salt layer is present at the site of the future diapir; (2) the gravitational load of the overlying layers a and b squeeze the salt sideways into a dome shape; extensional faults develop in the overlying competent layer a while deposition of layer c onlaps the sides of the dome; (3) with further doming, the salt breaks through the faulted layer a and spreads sideways into layer b to become a diapir; layers a, b and c are domed upwards and distorted into two asymmetrical synclinal structures at the sides of the diapir, while deposition of layer d takes place around its sides; (4) the diapir is now fully developed by spreading sideways into layer c, where it will now be more stable gravitationally. Based on Trusheim (1960).

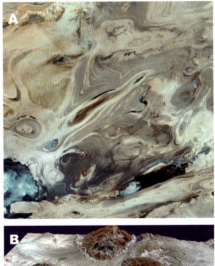

Figure 9.5 Salt domes in Iran. **A.** Satellite image of salt domes and related structures in the Dasht-e Kavir (Great Salt Desert), Central Iran; the salt domes are roughly circular in plan and possess a series of concentric rings; note the swirling pattern of the folded sedimentary layers between the domes. **B.** Salt glaciers flowing from salt domes in the Zagros Mountains; the glaciers are over 5 km long; the dark colour is due to clay covering the salt surface. Both images courtesy of NASA.

cylindrical in form, commonly 1–2 km in diameter, and then into a mushroom shape with a circular cross-section. In some places the salt may even pierce the surface, as in one of the salt diapirs of the Zagros range in Iran, where salt flows down the flanks of the dome, rather like a salt 'glacier' (Figure 9.5B). Where the initial salt body is in the form of an anticline, it will evolve into an elongate salt pillow. Any subsequent diapir will also be elongate and is termed a salt wall. Salt walls are often formed along steeply inclined faults.

The flow of the salt within these structures creates sets of folds and associated planar and linear fabrics that may be used to reconstruct the geometry of the whole structure. Plug-type diapirs possess internal structures whose flow directions are radial at the base of the structure, become parallel to the cylindrical stem of the diapir, and then spread out radially again in the upper mushroom (Figure 9.6A, B). Folds formed by the layers within the salt become tighter and more constricted as they approach the cylindrical stem of the diapir (Figure 9.6C, D) such that the fold axes, which initially would be straight or gently curved, become progressively more tightly curved, as shown in Figure 9.6C, individual folds assuming a rounded conical, or sheath-like shape (Figure 9.6F).

Structures also form in the strata surrounding the salt bodies (Figure 9.5A): the withdrawal of salt from a circular area around a salt dome will produce a syncline, which will be accentuated as the diapir breaks through to form a plug, as seen in Figure 9.4.4. Anticlinal structures formed above a salt dome are associated with extensional normal faults and graben. Structures around the margins of salt diapirs often form oil traps that are sealed by the impermeable salt – many oil fields in the Gulf of Mexico, for example, have been formed in this way.

Although salt flow can be quite fast compared to other rocks, salt diapirs can take many millions of years to evolve into their final (relatively stable) form. The numerous salt bodies of the north German plain arose from a Zechstein (upper Permian) evaporite layer which began to move upwards to form a set of diapirs during the Triassic. However, upward movement was still occurring in the early Cenozoic.

Granite diapirs and mantled gneiss domes

The granite diapirs discussed in the previous chapter were able to rise through their overburden because they were (largely) liquid as well as being less dense than their host rocks. However, the metamorphic core complexes of many orogenic belts contain dome-shaped or anticlinal bodies of largely granitic composition that represent older basement and which appear to have risen through their younger metamorphic cover. Many of these have been interpreted as diapirs that have moved up through the crust in a solid state and are therefore analogous to the salt diapirs just described. Such bodies are particularly common in Archaean granite–greenstone terrains, so-called

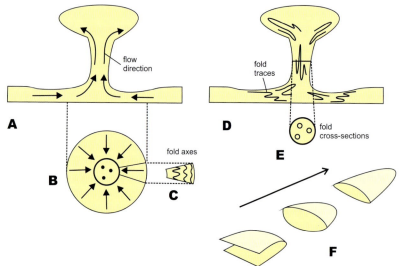

Figure 9.6 Diagrammatic representation of flow directions and fold patterns in a salt diapir. **A.** Cross-section of a salt diapir showing idealised pattern of flow. **B.** Plan view of base of diapir showing radial flow directions; flow is vertical in the central column. **C.** The folds are constricted as they move towards the central column, and the fold axes become progressively more curved. **D.** Cross-section of diapir showing idealised fold traces. **E.** Plan view through centre of column showing fold cross-sections. **F.** Progressive tightening and elongation of a fold with a curved axis.

because they are largely composed of granites and granite gneisses separated by areas of 'greenstone', which consist of low-grade metamorphic volcanic rocks and associated sediments such as greywackes (*see* Chapter 12).

Figure 9.7 illustrates the main features of an idealised granite–greenstone terrain. The granitic bodies occupy dome-shaped or rounded outcrops separated by the synclinal greenstone sequences, which tend to form cuspate margins projecting into the granites. The 'granites' are composed partly of older gneissose rocks, of predominantly granitic composition, and partly

of granites intruded into them; they often exhibit a margin-parallel foliation, similar to that shown by the intrusive granites described in Chapter 8. However, the greenstone sequences in many cases dip away from the granite margins and sometimes appear to be unconformable on them, suggesting a basement–cover relationship. In the late 1940s Eskola described such bodies in Finland as '**mantled gneiss domes**'.

The interpretation of these granitic bodies as gravity-driven domes and diapirs has been controversial. However, the density difference between granite and basalt – at about 0.2 Mg/m³ – is

not too different from that between salt and most other sedimentary rocks at around 0.3 Mg/m³; moreover, solid flow of rock at depth is required by considerations of isostatic behaviour (*see* Chapter 2). A series of famous experiments by Hans Ramberg in the 1960s using artificial materials in a centrifuge in order to carry out the experiments on a workable timescale, produced structures remarkably similar to those seen in nature, so there are good grounds for believing that solid flow of warm rocks at depth in the crust can explain the types of granitic structures that have just been described.

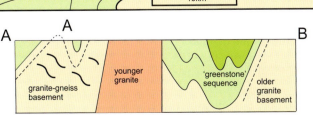

Figure 9.7 Diapirism of granite-gneiss basement. Idealised map and vertical cross-section along line A–B of an imaginary Archaean granite–greenstone terrain showing diapirs and domes of granitic basement 'intruding' into a synclinal cover consisting of lower and upper greenstone sequences. The basement consists partly of granitic gneiss and partly of intrusive granite. The older foliation in the basement is discordant with the greenstone cover, which is unconformable upon it. There is a newer foliation parallel to the margins of the up-domed basement. Both basement and cover are cut by a younger, undeformed granite.

10   Tectonic interpretation of orogenic belts

Most of the more interesting geological structures discussed in the previous chapters are found in orogenic belts, and the majority of structural studies have been carried out there. Such studies are important in understanding how orogenic belts can be explained by plate tectonic models and in linking structures seen on a small scale to their Earth-scale origins.

Principal tectonic units of the orogenic belt

The word 'orogenic' means 'mountain-forming', and in a geological context has come to be applied specifically to large-scale linear surface features exhibiting both lateral contraction and vertical uplift. The present-day orogenic belt system, described in Chapter 2, consists of two main strands: the circum-Pacific and Alpine–Himalayan, both of which vary widely along their lengths, and comprise both continental mountain ranges such as the American Cordillera, the Alps and the Himalayas, and partly submerged oceanic island arcs such as those in the north and west Pacific, Indonesia and the Caribbean (*see* Figure 2.2). In terms of the plate tectonic model, orogenic belts are explained as the sites of either plate subduction or continent–continent collision (*see* Chapter 3, Figure 3.10).

All orogenic belts contain a **central crystalline core** consisting of uplifted, highly deformed, metamorphic rocks and igneous (largely granitic) plutons.

This central core is flanked by regions of folded and thrust strata incorporating material deposited in elongate marginal basins, together with sedimentary sequences originally laid down in the **continental platform** regions or **forelands** bordering the orogenic belt (Figure 10.1A). This arrangement may be symmetric, as in the Alpine–Himalayan system with forelands on both sides, or asymmetric, as in the American Cordilleran belt, where the foreland is on one side only, the other side being oceanic plate. Asymmetric systems are the result of the subduction of oceanic plate along one side of the orogen, whereas symmetric systems result from continent–continent collision but retain features derived from their previous subduction history.

The geometry of an orogenic belt is shaped by its plate tectonic history: for example, the marginal regions will represent either an active or a passive plate margin. In the former case, they will display evidence of igneous activity consistent with a subduction zone (e.g. *see* Figure 3.10B); in the latter case, igneous activity will generally be absent. Orogenic belts situated on an active continental margin are thus typically asymmetric with many of the important planar structures, such as major thrusts, inclined towards the continent; this sense of asymmetry is referred to as the 'polarity' of the belt. All orogenic belts must commence as subduction zones, although these may be subsequently obscured by continental collision, as in Figure 3.10C. However, the subduction history is preserved within the orogenic belt and can be reconstructed by examining its structure.

The marginal zones
Foreland and foreland basin
A passive continental margin bordering a symmetric orogenic belt such as that shown in Figure 10.1A will display a characteristic sequence of sedimentary zones, as illustrated in Figure 10.1B. A traverse from the interior of the continent (the **foreland**) of the lower (subducting) plate towards the original site of the subduction zone will encounter first a sequence of shallow marine sediments, often dominated by carbonates, lying on continental basement: this is the **continental platform**. This zone is followed by the **foreland basin** (or **foredeep**) which results from the depression of the continental crust caused by the loading effect of the rising orogenic belt (Figure 10.1C) and contains a thick sequence of predominantly clastic sediments derived from the erosion of the main mountain range. This depositional basin is followed in turn by the foreland fold-thrust belt, which is characterised by a set of thrusts and related folds with a polarity corresponding to that of the subduction zone, that is, the thrusts dip away from the platform towards the core of the orogen.

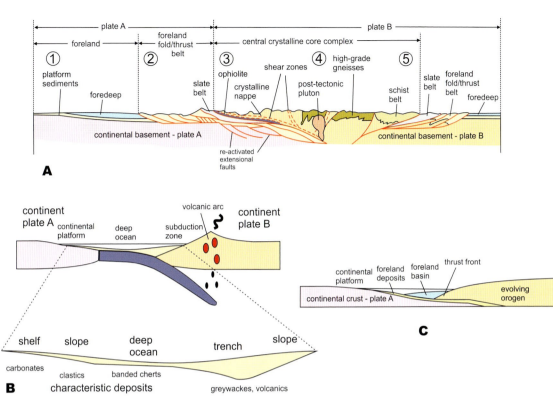

Figure 10.1 An idealised collisional orogenic belt. **A.** Cross-section, reading from left to right. (1) Plate A foreland: continental basement of plate A is overlain by a thin cover of platform sediments, which thicken towards the continental slope; they are overlain in turn by the foredeep or foreland basin, filled by sediments derived from the rising orogen. (2) The foreland fold-thrust belt: the outer part of plate A and its sedimentary cover are sliced up and telescoped by a set of thrusts and related folds; these are largely un-metamorphosed, but the outermost sediments may be of slate grade, forming the slate belt. (3) The central crystalline core complex: at the base of this tectonic unit is an ophiolite sheet that represents part of the oceanic crust and uppermost mantle scraped off the down-going slab and accreted to the base of the overlying continental plate. (4) Overlying the ophiolite sheet are a series of crystalline nappes (thrust sheets) consisting of high-grade, metamorphosed, sedimentary and plutonic igneous rocks affected by ductile folds and shear zones; these are initially directed towards the foreland but, towards the interior of the complex, they are affected by more upright structures; the complex is intruded by post-tectonic granite plutons. (5) Plate B foreland: the inner zones consist of a foreland fold-thrust belt directed towards the foreland, in a mirror image of the plate A margin; the highest unit consists of schist-grade metamorphics, succeeded in turn by a slate belt, un-metamorphosed foreland cover and a foredeep, overlying continental basement of plate B. Based on a diagram by Hatcher & Williams (1986). **B.** Schematic representation of the pre-collisional plate margin of plate A, showing the various depositional environments: shelf, slope, deep-ocean, trench and far slope, together with their characteristic deposits. **C.** Development of a foreland basin or foredeep. The weight of the advancing orogen depresses the thinned continental margin of plate A, causing a basin to form, which is filled by clastic sediments eroded from the rising mountains of the orogen.

The foreland fold-thrust belt

In order to understand the geometry of this belt, it is necessary to reconstruct the original make-up of the passive margin, as shown in Figure 10.1B. The depositional environments change ocean-ward as follows:

1. the **continental shelf** or platform, usually dominated by carbonates;
2. the **continental slope,** typified by coarse to fine clastic sediments deposited in deeper water, above the zone where the continental crust thins and is replaced by oceanic crust;
3. the **deep ocean** environment, characterised by thin-bedded cherts and mudstones;
4. the **ocean trench**, which is typically

filled by clastic material, especially greywackes, and volcanics derived from the volcanic arc on the upper plate; this region merges into:

5. the **continental slope** of the upper plate, beyond the subduction zone, which is also the site of thick clastic deposits.

The fold-thrust belt is a consequence of the convergence of the two opposing plates. Because the subducting slab dips beneath the upper plate, the leading edge of this plate acts like a snow-plough, scraping material off the top of the descending slab and thrusting it upwards and towards the approaching lower-plate continent (Figure 10.2). Thus the original lateral sequence of depositional environments as shown in Figure 10.1B has been telescoped into a comparatively narrow belt (Figure 10.1A).

As explained in Chapter 5, thrusts usually propagate forwards, away from the centre of the orogen, so that, in this case, successive thrust sheets carry material from progressively further away from the original site of the subduction zone. Thus the lowest and youngest thrust sheets contain platform sediments similar to those on the undeformed foreland; higher thrust sheets contain sediments deposited on the continental slope; above these are sheets containing material from the deep-ocean basin, and even higher sheets carry deposits from the trench and continental slope of the upper plate, which together make up what is known as the **accretionary prism** or **accretionary wedge**.

The accretionary prism

This consists of an accumulation of clastic sediments and volcanic debris occupying the trench and continental slope of an active continental margin, which is piled up in a series of folded and thrust slices (Figure 10.2). The accretionary prism is deformed progressively as subduction of the oceanic slab proceeds. It is underlain by a basal thrust, or detachment, above which the strata are scraped off and compressed. With continued convergence, the deformation front moves outwards towards the ocean; here thrusts and overfolds are directed oceanwards; the zone of maximum compression, where the sedimentary wedge is thickest, may be uplifted and the structures on the landward side directed towards the continent. At deep levels of the prism, temperatures and pressures are higher and structures more ductile. Higher and older thrust sheets may thus expose rocks that had been taken down to some depth on the subducting slab and now exhibit slate-grade metamorphism and more ductile deformation.

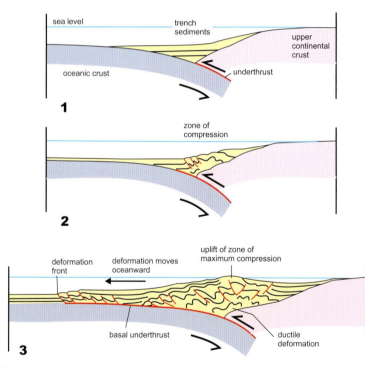

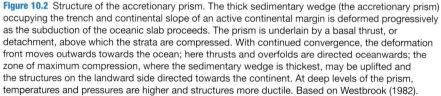

Figure 10.2 Structure of the accretionary prism. The thick sedimentary wedge (the accretionary prism) occupying the trench and continental slope of an active continental margin is deformed progressively as the subduction of the oceanic slab proceeds. The prism is underlain by a basal thrust, or detachment, above which the strata are compressed. With continued convergence, the deformation front moves outwards towards the ocean; here thrusts and overfolds are directed oceanwards; the zone of maximum compression, where the sedimentary wedge is thickest, may be uplifted and the structures on the landward side directed towards the continent. At deep levels of the prism, temperatures and pressures are higher and structures more ductile. Based on Westbrook (1982).

Because sedimentation continues as subduction and convergence proceed, younger sediments are less deformed than older, as shown in Figure 10.2.3.

Inversion

Some of the thrusts affecting the continental margin may represent re-activated normal faults formed as a result of the extensional event that originally created the passive continental margin (Figure 10.3). This re-activation is termed **inversion**, and occurs widely where compression of a piece of crust takes advantage of zones of weakness already created by a previous extensional episode.

Eu-geosynclines and mio-geosynclines

Before the plate tectonic theory was established, it was recognised from studies of many orogenic belts that there were two contrasting types of depositional basin: the **eu-geosyncline** and the **mio-geosyncline**, both containing thick sequences of predominantly clastic deposits of equivalent age to the much thinner platform sequences. The mio-geosyncline corresponds to the continental slope environment and the eu-geosyncline to the deep-ocean basin, ocean trench and the marginal depositional environment of the upper plate. A diagnostic characteristic of the eu-geosyncline was held to be the presence of volcanics and, importantly, ophiolites.

Ophiolites

The term **ophiolite** refers to a sequence of rock types that are interpreted as pieces of oceanic crust. A 'complete' ophiolite sequence would commence with ultramafic material assumed to be from the uppermost mantle, succeeded

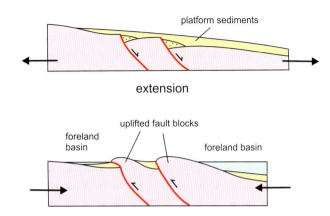

platform sediments

extension

uplifted fault blocks

foreland basin

foreland basin

compression

Figure 10.3 Inversion. Faults formed during the extensional period when the passive continental margin was created are re-used during orogenic compression by a reversal of the movement sense on them.

by a unit consisting of sheet-like mafic and ultramafic intrusions, followed in turn by a sheeted dyke complex, basalt pillow lavas and ocean-floor sediments. The origin of this sequence is due to the way that ocean crust is created, and is discussed in Chapter 3 (see Figure 3.9A). The ocean-floor assemblage contains the characteristic red cherts, which are thinly-bedded silicic deposits containing fossil radiolarian – **radiolarian cherts**; these are diagnostic of the deep ocean, since they are deposited beneath the level where carbonate can be preserved.

Complete sequences should be around 6 km thick at least, and do occur in some orogenic belts, but fragmentary sequences are quite common and occur in accretionary prisms where they have been scraped off the top of the downgoing slab. Some complete sequences occur in large thrust slices that are continuous for many hundreds of kilometres and may mark the boundary between two plates, as shown in Figure 10.1A, since they are the first slice to be scraped from the subducting slab, and therefore form the highest thrust unit

beneath the upper plate. The boundary between the plates is known as a **suture** or suture zone and is discussed below.

Outer zones of the opposing plate

Fold-thrust belts and foreland basins are also typically present on the opposite plate, as shown on the right-hand side of Figure 10.1A, but may have a different origin. The sense of asymmetry exhibited by that part of the orogen closest to the subduction zone (i.e. the fold-thrust belt just described) does not appear to be transmitted through the orogen to the far side. The convergence between the two plates, and the consequent uplift of the central core complex, have the effect of imposing the opposite sense of asymmetry to the marginal zones on the other plate. Thus it seems that all orogenic belts are bordered by thrust belts that combine to squeeze the orogen outwards towards the bordering plates.

The central crystalline core complex

The central core of the orogenic belt consists essentially of originally deep-seated crystalline rocks that have been raised up and unroofed by erosion to expose a complex region of deformed high-grade metamorphic rocks and igneous plutons. Faults in these deep-seated rocks are replaced by shear zones, and folds display the effects of ductile deformation.

Where the upper plate consists of continental crust (as in Figure 10.1A), the rocks lying immediately above the suture, and which have been thrust up over the units of the foreland thrust belt, may represent the lowermost part of the continental crust; in that case, they will display the effects of high-pressure metamorphism, including granulite facies and, especially, **eclogite facies**. The latter are particularly important, since they signify a high-pressure, moderate-temperature regime corresponding to that of the relatively cool, but very deep, regions above the subducting slab. These metamorphic rocks are predominantly gneisses of both meta-sedimentary and meta-igneous origin, the latter being mostly of broadly granitic composition.

Granites, granodiorites and diorites are the characteristic products of subduction-related igneous activity, and represent the deep-seated counterparts of volcanic arcs. Such rocks are usually deformed by the effects of plate collision, but are intruded by a later set of plutons (typically granitic), often in the form of large batholiths. These have resulted from the melting of deep-crustal material as a result of the rise in temperature brought about by crustal thickening, and are known as **post-tectonic** plutons in contrast to the **pre-tectonic** plutons that are generated by the subduction process.

Structures of the core zone

Structures in the central part of the orogenic belt may be very complex and difficult to interpret. The oldest will represent deformation that occurred during the formation of the continental basement and will throw no light on the orogenic process itself. However, structures affecting the pre-tectonic plutons and any sedimentary or volcanic rocks deposited on the older basement will provide a guide as to how the orogen developed.

The first structures to form are generally large recumbent or inclined isoclinal folds and related shear zones that reflect the asymmetry of the fold-thrust belt of the foreland, but involve higher-temperature schists and gneisses. The ductile nature of these structures is due to the higher temperature within the central part of the orogen, brought about partly by the heating effect of the igneous intrusions and partly by the crustal thickening. These early folds are typically refolded by more upright folds that result from the sub-horizontal compression caused by the convergence of the opposing plates. In some cases, in favourable lithologies, several generations of younger folds with their accompanying fabrics can be identified, and these generally reflect a continuation of lateral compression within the orogen.

The deformation just described results in both shortening and thickening of the crust in the core zone to the extent of doubling or even tripling the original crustal thickness. Seismic studies of some orogenic belts have revealed **Moho** depths of up to 80 km.

Gravitational spreading and channel flow

Because of the increased ductility of the warm lower crust of a thickened orogen, this region is more susceptible to gravity-induced flow, or **gravitational spreading**, as described in the previous chapter (see Figure 9.3). Under these conditions, the continental lithosphere of the orogen acts like a sandwich with two stronger layers – the upper crust and the lower crust/uppermost mantle – separating a ductile layer which is subject to lateral flow and is squeezed towards the sides of the orogenic belt (Figure 10.4). This process has been termed **channel flow** and is a possible explanation for many of the ductile structures seen in orogenic belts where thrust sheets containing medium- to high-grade gneisses are bounded by shear zones with opposite shear senses, and are both underlain and overlain by stronger, less ductile material.

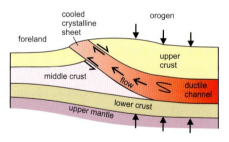

Figure 10.4 Channel flow. In a developing orogen, the middle crust becomes heated and partially melted by the injection of igneous plutons and by depression to warmer depths due to crustal thickening, and forms a ductile channel. The gravitational effect of the thickened crust squeezes this more ductile middle-crustal material, enabling it to flow laterally and escape to the surface as it cools; there it forms a crystalline sheet bounded by upper and lower shear zones with opposite senses of shear.

Terranes and micro-plates: compound orogenic belts

Only the simplest orogenic belts consist of two continental plates separated by a single suture zone. Many, on the other hand, are made up of several distinct pieces of continental crust that have become welded together, each separated from its neighbour by a **suture zone**. Figure 10.5 shows one such example, where two plates enclose a smaller piece of crust that could be referred to as a **micro-plate**. Geologists working in the northern American Cordillera have distinguished a number of such crustal units there, referring to them as **terranes**. They have developed several geological criteria for distinguishing terranes: they must exhibit a different structural history from neighbouring regions and ideally contain palaeomagnetic or fossil evidence of derivation from some distance away from their neighbours. In the North American case, some of these terranes represent pieces of crust that have broken away from another continent and migrated across the ancestral Pacific Ocean to collide with the North American plate once the intervening oceanic lithosphere has been subducted. Such a collision is known as a **docking** event. Other terranes have been moved to their present positions along major transform faults.

Terranes are now widely recognised as an important constituent of many orogenic belts; they are of three main types: continental micro-plates that have separated from another continent; volcanic island arcs; and pieces of unusually thick oceanic crust known as **oceanic plateaux**. The process of **back-arc spreading**, discussed in Chapter 3

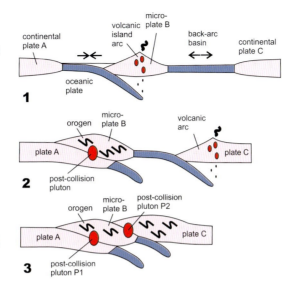

Figure 10.5 Terrane amalgamation. Cartoon showing how a compound orogenic belt composed of more than two colliding plates (cross-section 3) can be understood in terms of the sequence of collisions of its constituent parts. Thus in cross-section 1, two continental plates A and C are about to collide with an intervening volcanic arc (or micro-plate) B; in (2), plate A has collided with micro-plate B, while B is still subducting beneath C; in (3), the combined plates A and B have now collided with C to form a compound orogen consisting of three terranes. The sequence of docking of the terranes is revealed by comparing the age of the post-collision pluton P1 with that of the younger post-collision pluton P2.

(*see* Figure 3.11) is an important source of micro-plates. This seems to occur when the upper plate and the trench are moving away from each other. The resulting extensional stress causes thinning of the continental lithosphere, and ultimately the creation of new oceanic lithosphere behind the volcanic arc.

Compound orogenic belts consisting of several amalgamated terranes, which may have docked with the main plate at different times, inevitably display a more complex geometry and tectonic history than one formed by a single collision event (Figure 10.5). Each separate collision will involve a subduction process followed by a collision event, both with their accompanying structures, and it is necessary to use careful analysis and sophisticated dating methods to separate them. In Figure 10.5, the sequence of docking is revealed by comparing the dates of the two post-collisional plutons P1 and P2: since P1 is older than P2, plate A

must have docked with micro-plate B before the latter docked with plate C.

Oblique convergence

There are few presently active orogenic belts where the convergence direction is exactly perpendicular to the trend of the belt. A glance back at the plate movement vectors in Figure 3.7 shows that in many cases the convergence direction across the plate boundary is highly oblique: see, for example, the Pacific–North American boundary and the northern Indonesian sector of the Indian–Asian boundary. It is usually easier for structures to form parallel to major pre-existing crustal features, such as the continental margin itself. Therefore, oblique convergence has the result of dividing the oblique compressional stress across the boundary into two sets: one set, acting at right angles to the boundary, producing folds and thrusts that trend parallel to it, and a second set, acting parallel to the boundary, producing strike-slip faults that are also parallel to the boundary (Figure 10.6). This process is termed **partitioning**; the compressive stresses are said to be **partitioned** into the two sets.

Major strike-slip movements may thus take place along any convenient steep fault near the margins of an orogenic belt subjected to oblique convergence, or on any suitably oriented terrane boundaries within the belt. Such a process has the unfortunate result that adjacent fault blocks or terranes may not exhibit the same geological history, which may cause confusion in the interpretation of the evolution of the orogenic belt as a whole. Examples of this are discussed in the following chapter.

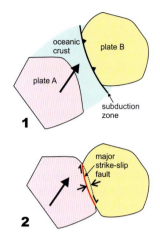

Figure 10.6 Oblique collision. (1) Plate A approaches plate B obliquely with respect to the plate B margin; in (2), the oblique convergence is partitioned into a compression acting across the margin, producing margin-parallel folds and/or thrusts, and a major dextral strike-slip fault parallel to the margin.

11   Examples of modern orogenic belts

The present-day system of orogenic belts is briefly described in Chapter 2 (see Figure 2.2). We shall take, as examples of modern orogenic belts, two sectors of the Alpine–Himalayan orogenic belt that are comparatively well known: the central Himalayas and the Western (French-Swiss) Alps to represent the continent–continent collision type of belt, and the Canadian sector of the North American Cordilleran system as an example of the continental margin type. The Cordilleran system runs along the western side of the Americas from Alaska to southern Chile. Active subduction in the North American Cordillera is confined to two comparatively short sectors: north of the Aleutian trench in Alaska, and west of the Cascades volcanic arc in the north-western USA (see Figure 3.6B). The remainder of this system, including the Canadian sector, lies adjacent to major transform faults and contains multiple displaced terranes.

The layout of the modern orogenic belt network is a result of plate movements during the period from the Jurassic to the present day and is summarised in Figure 11.1.

The Himalayan belt

The Himalayan orogenic belt is part of the long and complex system of orogenic belts that extends from the western end of the Mediterranean to Burma (Myanmar) in the east (*see* Figure 2.2) and includes the great chains of the Alps, Carpathians, Caucasus and Zagros ranges, and, most prominent of all, the Himalayas. The Himalayan sector itself is over 2,500 km long and is draped around the northern perimeter of the Indian continent, which projects into it at its western and eastern ends (Figure 11.2). It contains several of the world's highest mountains, including Mount Everest, and is bordered to the north by the high Tibetan plateau, with a mean elevation of over 5,000 m. The western (Pakistan) sector is known as the Karakorum.

The Himalayan belt is the result of the collision of the Indian continent with Central Asia (Figures 11.1, 2), which is itself an amalgamation of several continental blocks including North and South Tibet and Tarim to the north, and the South-China and North-China blocks to the north-east, all of which joined the Siberian core of Asia during the Mesozoic. The sutures between these blocks represent lines of weakness within the Asian continent, which were exploited during the Indian–Asian collision, and along which renewed activity in the form of thrusting or strike-slip faulting took place.

Our knowledge of the history of the Himalayan belt is assisted by detailed information from the Indian Ocean magnetic-stripe data (*see* Chapter 3; e.g. Figure 3.5 and 3.6), from which it is deduced that India, having detached from the Gondwana supercontinent during the Cretaceous, probably reached the margin of the Asian continent at around 50 Ma ago. Prior to

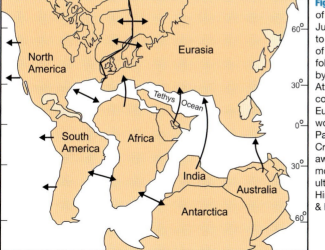

Figure 11.1 The break-up of Pangaea. During the Jurassic, Pangaea started to break up by the opening of the central Atlantic, followed in the Cretaceous by the opening of the South Atlantic and consequential convergence of Africa and Europe. The north Atlantic would not open until the Palaeocene. Also in the Cretaceous, India broke away from Gondwana and moved north towards Asia, ultimately to create the Himalayas. Based on Smith & Briden (1977) map 7.

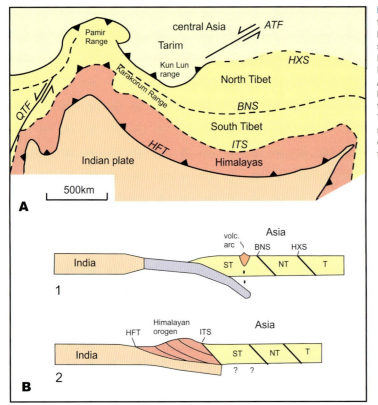

Figure 11.2 The Himalayan orogenic belt. **A.** Simplified map showing the main tectonic units and structural lineaments of the Himalayan belt; the main mountain belt is shown in red, the Asian plate in shades of yellow; the Indian plate in brown. HXS, Hoh Xil suture; BNS, Bangong– Nujiang suture; ITS, Indus–Tsangpo suture; HFT, Himalayan frontal thrust; QTF, Quetta–Chaman strike-slip fault; ATF, Altyn Tagh strike-slip fault. **B.** Cartoon cross-sections illustrating how the Himalayan belt evolved from the subduction phase (1) to the collision phase (2). ST, South Tibet block; NT, North Tibet block; T, Tarim block; other symbols as (A). Note sutures BNS and HXS marking the sites of earlier subduction–collision events. The extent of Indian plate beneath Tibet is uncertain but probably extends to the BNS suture (see Figure 11.4).

this event, subduction of the oceanic part of the Indian plate had been taking place beneath Asia from Cretaceous times. The climax of the collision event resulting in the uplift of the Himalayan range itself occurred in the Miocene, around 20 Ma ago, although uplift and thrusting along the southern border continues today.

The central Himalayan sector

This sector of the Himalayan belt (Figure 11.2) forms an arc, 1,750 km long and 250 km across, extending from north-western India, through Nepal, to Bhutan, and includes southernmost Tibet. It is bounded on the north by the

Indus–Tsangpo–Yarlung Suture (ITS) which marks the junction between the Indian and Asian plates. In the central sector the suture is offset by a thrust – the **Renbu–Zedong Thrust** (RZT). The southern margin of the belt is defined by another thrust, the **Himalayan Frontal Thrust** (HFT), which marks the edge of the fold-thrust belt on the Indian foreland. The central Himalayan belt consists of the following four separate tectonic units, traversing from south to north (Figures 11.2, 11.3).

◆ 1. The **foreland thrust belt**. This unit consists of south-directed fold-thrust sheets involving the **Siwalik**

Formation, which consists of un-metamorphosed clastic sediments formed in the **foredeep basin** and derived from the rising Himalayan mountains.

◆ 2. The **Lesser Himalayan schist belt**. This unit consists of slates and schists derived from Mid-Proterozoic clastic sediments, originally laid down on the Indian passive margin and deformed into south-directed fold-thrust packages. The southern boundary is marked by the **Main Boundary thrust** (MBT).

◆ 3. The **Greater Himalayan crystalline complex**. This unit consists of high-grade schists and gneisses derived from Late Proterozoic clastics, also from the Indian passive margin. The metamorphic grade is inverted, becoming greater upwards, from chlorite to sillimanite at the upper margin of the belt. Near the upper margin is a zone of syn-orogenic granite intrusions with a generally lensoid shape. The southern boundary of the crystalline complex is formed by the **Main Central thrust** (MCT), which is actually a ductile shear zone several kilometres wide.

◆ 4. **The Tethyan shelf belt**. This unit consists of largely un-metamorphosed

Cambrian to Eocene marine strata, mainly carbonates and shales, originally laid down on the Indian continental shelf. The lower boundary of this belt is a north-dipping normal fault – the **South Tibetan detachment** (STD).

The northern margin of the central Himalayan sector is the south-dipping **Indus–Tsangpo–Yarlung suture** zone (ITS), which contains an ophiolite complex of Cretaceous to early Tertiary age. In the central sector of the orogen, the suture is cut by the south-dipping **Renbu-Zedong thrust** (RZT).

The Asian plate
The geology north of the suture is less well known than that of the main Himalayan belt. Immediately north of the suture is an elongate granite batholith (the **Gangdese batholith**), which represents part of a volcanic arc resulting from the subduction of Indian oceanic lithosphere in the early Cenozoic. The Asian plate has been less obviously affected by the Himalayan compressional deformation, which is partly concentrated along the suture zones between the microplates (e.g. the South Tibet, North Tibet, and Tarim blocks) that had previously accreted to Central Asia and partly taken up by movements along a network of faults. Numerous north–south oriented graben systems indicate significant E–W extension and this, coupled with movements on a **conjugate** set of strike-slip faults, has been interpreted to indicate that much of the north–south convergence between India and Asia has been accommodated by the sideways extrusion of Asian crust.

Repeated precise GPS measurements have enabled accurate movement vectors across the orogen to be calculated (Figure 11.4); these range generally between 5 and 15 mm/year, confirm the lateral extrusion model, and also indicate a gradual diminution of flow velocity northwards – there is thus no clearly defined northern margin to the deformation resulting from the collision as there is in the south. GPS and **InSAR**-derived data on slip rates along strike-slip faults are typically in the range 5–15 mm/yr; interestingly, the slip rates on the very large strike-slip faults such as the Altyn Tagh and Karakorum, formerly believed to have accommodated the bulk of the N–S compression, are little different from the others.

This pattern of distributed deformation, achieved mainly by faulting, applies only to the upper (seismogenic) crust. Beneath this, the middle crust, being warmer and more ductile (*see* below), will have deformed in a more continuous manner, employing shear zones rather than discrete faults.

Orogenic history
The three northern units of the central Himalayan belt all represent packages of material scraped off the top of the continental Indian plate as it descended beneath the Asian plate after the initial collision and emplaced as thrust sheets. This process is illustrated schematically in Figure 11.3C, D. It is assumed that the thrust sheets developed by propagating forwards with a ramp-flat geometry as described in Chapter 5 (*see* Figure 5.7). As each sheet ramped up, the orogen would shorten and thicken, although erosion would continuously remove material from the roof of the structure.

Two alternative explanations have been put forward to explain the juxtaposition of the high-grade gneisses of the Greater Himalayan complex (GHC) and the Tethyan shelf sediments (TSS) above them, separated by the South Tibetan detachment (Figure 11.3B). The 'orthodox' ramp-flat thrust model, as indicated above, suggests that successive forward-propagating thrusts would cause the GHC and its TSS cover to be arched up and subjected to erosion. Gravitational spreading then caused the TSS to slide down to the north, exposing the higher-grade rocks of the GHC.

The other explanation relies on the **channel-flow mechanism** (*see* Figure 10.4). In this model, the GHC is regarded as a hot, partially molten piece of Indian crust which has flowed under gravitational pressure, from its original position beneath the Asian plate, upwards to the surface between the TSS and the Lesser Himalayan schists beneath. It is possible that both mechanisms may play a part.

The 5 km-high Tibetan plateau, underlain by 70–90 km thick crust, is believed from seismic evidence to be supported by relatively cool and strong lower crust and lithospheric mantle of the Indian plate (Figure 11.3D). This broad, semi-rigid slab has underthrust the Asian plate for a distance of 200–300 km up to the BNS suture and is the main driver for the deformation of the orogen. The movement has absorbed 36–40 mm/yr of relative motion between India and Asia since the initial collision (i.e. ~2000 km of relative motion) of which an estimated 800–1000 km is taken up by shortening of Indian upper crust in the Himalayas and a further shortening of possibly ~600 km (accomplished mainly by

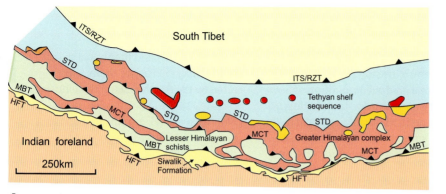

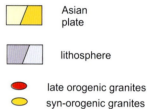

Indian plate

■ Cambrian-Eocene Tethyan shelf sequence

■ Greater Himalayan crystalline complex: Late Proterozoic clastics

■ Lesser Himalayan schists: Mid-Proterozoic clastics

■ Siwalik Formation: foredeep basin

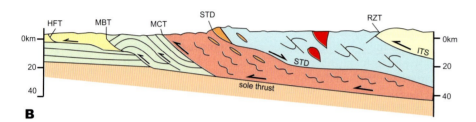

Asian plate

lithosphere

● late orogenic granites

● syn-orogenic granites

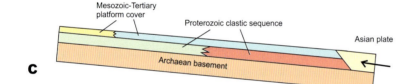

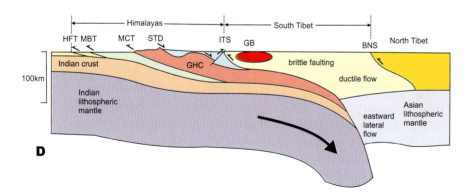

Figure 11.3 A, B. Map (A) and cross-section (B) of the main Himalayan mountain belt between the South Tibet block and the Indian foreland, showing the main tectonic units, from north to south: Cambrian to Eocene Tethyan shelf unit; Greater Himalayan crystalline complex; Lesser Himalayan schists; Siwalik foredeep basin; the thrusts separating these units are: RZT, Renbu-Zedong thrust (Indus–Tsangpo suture); STD, South Tibetan detachment; MCT, Main Central thrust; MBT, Main Boundary thrust; HFT, Himalayan Frontal thrust. Note the syn-orogenic granites near the STD boundary and the late orogenic granites within the Tethyan sequence. Modified from Harrison (2006). **C.** Simplified cartoon cross-section illustrating how the general structure of the orogen may be derived from an initial platform cover resting on Indian Archaean basement by pushing from the leading edge of the Asian plate, acting as a kind of plunger, along the top of the basement. Note that, for simplicity, the two Proterozoic formations are regarded here as laterally equivalent and differing only in their degree of metamorphism. Also, the platform cover on the foreland (Siwalik formation) is approximately time-equivalent to the Tethyan sequence in the northern part of the belt. D. Simplified cross-section through the central sector of the orogen to illustrate the contrasting arrangement of crust and lithospheric mantle between the Indian and Asian plates, based on Searle et al. (2011).

lateral extrusion) of the South Tibet block. Relative movement continues into the Asian plate north of the BNS suture, but at a slower rate (Figure 11.4).

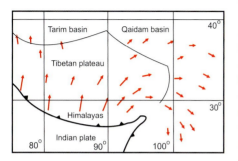

Figure 11.4 Velocity vectors (red arrows) relative to stable Eurasian plate derived from repeat GPS measurements. Note that the velocities decrease northwards and indicate flow is progressively north-eastwards then eastwards and south-eastwards in the eastern part of the sector. Based on Gan et al. (2007).

Orogenic strain

Rough estimates of bulk two-dimensional shortening strain across the central part of the orogen, using the displacement data recorded above, are ~78% for the Himalayas (i.e. the Indian upper-middle crust) and ~60% for the South Tibet block (the Asian crust). The shortening across the Himalayas is accomplished mainly by crustal thickening, much of which has been removed by erosion. The present Moho depth descends from ~40 km at the orogenic front to ~70 km beneath the ITS suture and remains around that level across the South Tibet block (see Figure 11.3D). The thickening of the Asian crust is thought to have taken place mainly during the pre-collisional subduction phase; during the collision phase, the warm middle and lower crust would have deformed partly by lateral ductile flow towards the east, corresponding to the sideways block faulting of the upper crust.

The Western Alps belt

The French-Swiss sector of the European Alps is the best known of all the young orogenic belts and has been studied intensively by some of the most famous geologists in the world since the mid-nineteenth century. Unfortunately, however, it is extremely complex – more so than the Himalayas – and therefore cannot, in some respects, be regarded as typical of orogenic belts in general. Several reasons for this are apparent from Figure 11.5, which shows in a simplified way the general shape of the western end of the Alpine–Himalayan orogenic belt system, from Gibraltar in the west to the Carpathians in the north-east and Greece in the south-east. We can see from this map that the orogenic belt occupies both sides of the western Mediterranean in southern Spain and North Africa, then swings up through Sicily and the Apennines. In south-eastern France, it curves around

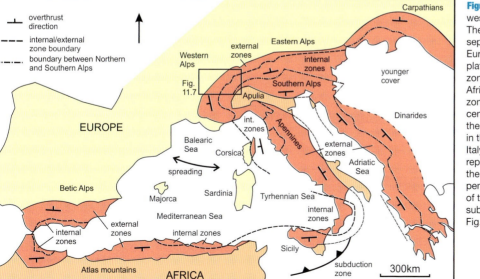

Figure 11.5 Alpine orogenic belts of the western and central Mediterranean. These orogenic belts, shown in red, separate continental crust of the European plate from that of the African plate. The belts are divided into external zones overthrust onto European or African plates respectively and internal zones, representing the uplifted central crystalline complexes. Note the promontory of African plate crust in the Adriatic, extending into northern Italy. The Balearic and Tyrrhenian seas represent new ocean crust created by the anticlockwise rotation of the Italian peninsula. In the south-eastern portion of the map, Neotethys ocean crust is subducting beneath Sicily. Based on Fig. 1 in Coward & Dietrich (1989).

through nearly 180°, describing a great arc through eastern France and Switzerland until, in the Austrian sector, it trends roughly east–west. The chain then divides into two branches, one continuing down through the **Dinarides** to Greece, and the other forming the great arc of the Carpathians.

The northern branch of the system includes tectonic units that are marginal to the European plate and have been overthrust onto the European foreland. This branch includes the **Betic Alps** in southern Spain, the Western and Eastern Alps proper, and the Carpathians. The 'southern' branch consists of the coastal sector of the **Atlas** mountains,

the **Apennines** of Italy, the **Southern Alps**, and the Dinarides; these are marginal to, and have been overthrust onto, the Apulian micro-continent (**Apulia**), which originally belonged to the African plate. Part of the complexity of shape is due to the fact that much of the western Mediterranean (the Balearic and Tyrhennian Seas), consists of 'young' ocean basins that have opened up during the last 10 million years and have resulted in the Italian peninsula being rotated through nearly 90°.

Plate movements

An important reason for the complexity of the Alpine belt is the movement history of the neighbouring plates, as summarised in Figure 11.1. The central Atlantic Ocean opened during the Jurassic Period, while Europe was still attached to America. This had the effect of moving Africa eastwards with respect to Europe. However, when the South Atlantic opened in the Cretaceous, the African plate changed direction with respect to Europe and rotated anti-clockwise, so as to move north, then north-west, towards Europe. Finally, as the North Atlantic opened during the Tertiary, Europe moved eastwards, rotating anti-clockwise in relation to Africa. These movements resulted in the following tectonic events (Figure 11.6).

◆ 1. The separation of the **Apulian microplate** from Africa (in the Cretaceous) due to the formation of the **Neo-Tethys ocean** as Africa moved eastwards.

◆ 2. Closure of the **Tethys Ocean**, and subduction of Tethyan crust beneath Apulia as Africa converged on Europe.

◆ 3. Collision of Apulia with the **Pennine terrane**.

◆ 4. Subduction beneath the Pennine terrane and collision of the combined Apulia–Pennine terranes with Europe in the main Alpine event (Miocene).

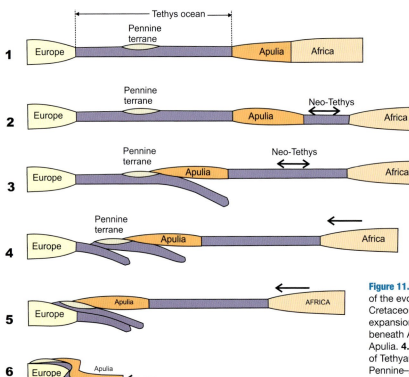

Figure 11.6 Cartoon cross-sections showing a possible interpretation of the evolution of the central Alps. **1.** Pre-Cretaceous. **2.** Mid-Cretaceous: opening of Neo-Tethys ocean. **3.** Late Cretaceous: expansion of Neo-Tethys causes subduction of Tethyan crust beneath Apulia and collision between the Pennine terrane and Apulia. **4.** Eocene: northward movement of Africa causes subduction of Tethyan crust of the European plate beneath the combined Pennine–Apulian microplate. **5.** Oligocene: further northward movement of Africa causes collision of the Pennine–Apulian microplate with Europe. **6.** Miocene: further convergence causes backthrusting of the thickened orogen onto Apulia.

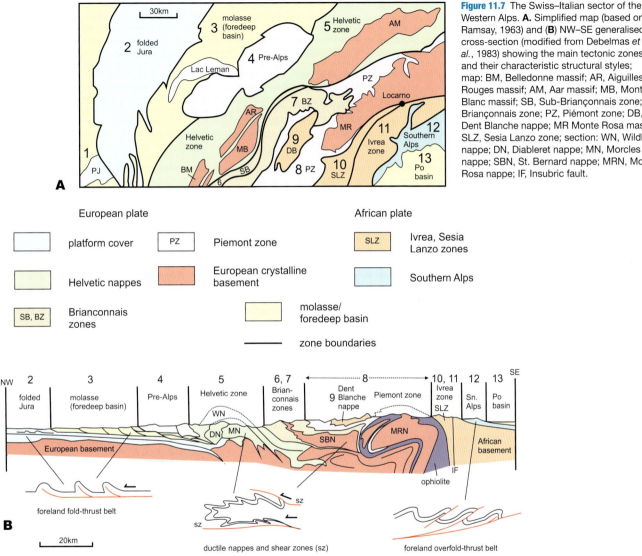

Figure 11.7 The Swiss–Italian sector of the Western Alps. **A.** Simplified map (based on Ramsay, 1963) and (**B**) NW–SE generalised cross-section (modified from Debelmas *et al.*, 1983) showing the main tectonic zones and their characteristic structural styles; map: BM, Belledonne massif; AR, Aiguilles Rouges massif; AM, Aar massif; MB, Mont Blanc massif; SB, Sub-Briançonnais zone; BZ, Briançonnais zone; PZ, Piémont zone; DB, Dent Blanche nappe; MR Monte Rosa massif; SLZ, Sesia Lanzo zone; section: WN, Wildhorn nappe; DN, Diableret nappe; MN, Morcles nappe; SBN, St. Bernard nappe; MRN, Monte Rosa nappe; IF, Insubric fault.

◆ 5. The opening of new oceanic crust in the western Mediterranean (the Balearic and Tyrhennian basins after the main Alpine event (Pliocene to Present) as Europe rotates anticlockwise.

Tectonic framework of the Western Alps

Figure 11.7 shows a simplified map of part of the Western Alps (*see* Figure 11.5 for location) in the NE–SW-trending section encompassing the Swiss Alps, which is the most complex tectonically and contains most of the highest peaks. This region is usually divided into 13 tectonic zones; these are as follows, described from west to east.

◆ 1–3. **The foreland**. Mesozoic platform sediments rest on the margin of the European plate; they are unfolded in zone 1 (the **Plateau Jura**) and folded in zone 2 (the **Folded Jura**). Zone 2 represents the outermost part of the foreland fold-thrust belt but is separated from the main part of the fold-thrust belt by zone 3, the **foredeep basin**, containing non-marine (continental) clastic sediments (**molasse**) derived from the evolving mountain chain.

◆ 4. **The Pre-Alps**. This zone is a **klippe** (thrust outlier) of rocks belonging to the Piémont zone (8).

◆ 5. **The Helvetic zone**. This consists of a set of complex fold **nappes** underlain by ductile thrusts or shear zones, and cored by crystalline European basement. These structures are directed north-westwards, towards the foreland. The sedimentary cover consists of Mesozoic platform sediments that can be correlated with those of the foreland, and represent the thinned margin of the European continent. The three main nappes shown in the cross-section are the **Morcles**, **Dent Blanche** and **Wildhorn** nappes. These nappes form many of the high mountains of the Swiss Alps.

◆ Zones 2–5 are referred to as the '**external zones**' of the orogenic belt and together make up the **foreland** fold-thrust belt.

◆ 6–7. **The Sub-Briançonnais** and **Briançonnais** zones. These zones consist of material thought to represent a small continental terrane, known as the **Pennine terrane**, situated on Tethys Ocean crust (*see* Figure 11.6) together with marine clastic sediments thought to have formed in a trench or back-arc basin. The Mesozoic sedimentary sequence is quite different from that of the foreland and is underlain by crystalline basement, which forms the core of the large **St. Bernard nappe**. The overfolds in this zone are in the form of a fan, directed towards the foreland in the north-west, like those of the underlying zones, but directed towards the south-east at the opposite side of the zone.

◆ 8. **The Piémont zone**. This zone contains ophiolites and typical ocean-floor sediments representing pieces of Tethys Ocean crust and upper mantle, which has been thrust over the continental crystalline basement that forms the core of the **Monte Rosa nappe**. The presence of ophiolites here is taken to indicate the presence of a suture between this zone and the overlying Dent Blanche nappe.

◆ 9–10. **The Sesia Lanzo zone** and the **Dent Blanche nappe**. The Dent Blanche nappe is a klippe of crystalline basement of African parentage belonging to the Apulian microplate. It is considered to have rooted in the Sesia Lanzo zone, which forms the northern margin of Apulia.

◆ 11. **The Ivrea zone**. This zone consists of crystalline basement similar to that of the Sesia Lanzo zone, but separated from it by a major steep fault, the **Insubric fault**, along which strike-slip movement took place during the latter stages of the orogeny.

◆ 12. **The Southern Alps**. This zone, which includes the **Dolomites** range, consists of a SE–directed fold-thrust belt affecting Mesozoic platform sediments belonging to the African plate.

◆ 13. **The Po basin**. This zone consists of non-marine clastic sediments derived from the rising Alps and deposited in a foredeep basin on the Apulian plate.

Structures

The structures of the outermost zones, the folded Jura and the Southern Alps, exhibit typical fold-thrust geometry (Figure 11.7B), with outward-directed thrusts and thrust-generated overfolds linked to a sole thrust along a weak stratigraphic horizon, in the same way as was illustrated in the outer zones of the Himalayas. The nappes of the Helvetic zone, having originated at greater depths, are more ductile, but still have an overall, foreland-directed, overthrust sense of movement (Figure 11.7B). The more complex nature of the folding in the Briançonnais and Piémont zones, illustrated in Figure 11.7B, has been attributed to SSE-directed '**back-thrusting**'. This has had the effect of moving the Piémont zone upwards and backwards, towards the Apulian plate, thus raising the metamorphosed interior of the orogen to a higher level and isolating the Dent Blanche nappe (*see* cross-section 6 of Figure 11.6). This back-thrusting may be attributable to continued convergence between the opposing continental plates at a time when further overthrusting was inhibited by the already greatly thickened orogen.

Detailed analysis of the ductile structures of the Piémont zone indicates that the movement direction changed through time in an anticlockwise sense, suggesting a link with the anticlockwise change in the plate convergence direction noted above. A similar pattern has been observed in the outer nappes of the Helvetic zone, with the outer, younger thrusts showing a more north-westerly movement direction. Gravity gliding is thought to be responsible for the isolated position of the Pre-Alps and for certain nappes in the Helvetic zone (e.g. *see* Chapter 9 and Figure 9.2).

The ophiolites of the Sesia Lanzo zone exhibit very high-pressure metamorphism, indicating that they had been subducted to considerable depths before being upthrust to their present level.

Attempts have been made to estimate the total amount of shortening across the Alpine belt by restoring the movements along the various thrusts. Accurate measurement is impossible because of the complexity of the structures, especially in the interior zones, but it is likely that the shortening in the continental crust has been in excess of 250 km across the belt, which is now ~300 km in width. This compares with a crustal thickening in the orogen to over 50 km, the present depth of the Moho there.

Timing of the orogeny
The complex shape of both plate margins, together with the change in the convergence direction, have resulted in regional variation in the commencement of the Alpine orogeny. In the sector described above, crustal deformation seems to have

commenced during the Cretaceous in the form of thrusting onto the margin of the Apulian micro-plate; however, collision between that plate and the main European plate probably did not occur until the late Eocene or early **Oligocene**. Crustal shortening and orogen creation then continued through the Oligocene, reaching its climax in the late Oligocene, about 25 Ma ago. The late back-thrusting phase has been attributed to the Miocene period. However, some convergent movement and uplift still continue today.

The Canadian sector of the North American Cordilleran belt
The Canadian Cordillera extends from the Yukon Territory in the north to the northern border of the USA in the south, and includes the coastal region of southern Alaska (Figure 11.8A). It is approximately 2,000 km long and 1,000 km wide. The eastern part of the mountain range consists of a fold-thrust belt directed towards the eastern foreland of the North American plate (Figure 11.8B). The sedimentary cover in this belt consists of Palaeozoic to Mesozoic platform and continental slope deposits resting on the Precambrian basement of the **Laurentian shield**. These strata record a passive margin of the North American continent which existed through the Palaeozoic until, in the Mesozoic, subduction of the **Farallon plate** (*see* Figure 3.6B) commenced at its western margin.

The main part of the orogen is composed of a number of displaced

terranes (*see* Chapter 10), which have accreted to the North American plate during the Mesozoic. The main collision event took place during the Cretaceous period. The terranes represent a mixture of crystalline metamorphic complexes, volcanic arc material, arc–trench deposits of various ages, and ophiolite units containing oceanic sediments and ultramafic rocks (Figure 11.8A, B). These in some cases can be shown to have originated far from their present position and are believed to have been transported across the ancestral Pacific Ocean on the Farallon plate and scraped off the subducting slab.

Several of the terranes, including the **Wrangellia** and **Alexander** terranes, accreted together first during the Jurassic, before colliding with the continent in the Cretaceous. This collision was responsible for much of the deformation in the orogen. The later history of the orogen was marked by lateral movements along major strike-slip faults, parallel to the transform fault that forms the present Pacific–American plate boundary. These movements slid the various terranes northwards relative to the eastern part of the orogen. Late-orogenic collapse due to gravitational spreading has resulted in extensional faulting that has reversed the movement on some of the earlier thrust faults. The north-western part of the orogen consists of a Cenozoic accretionary prism, and parts of the coastal belt are still tectonically active.

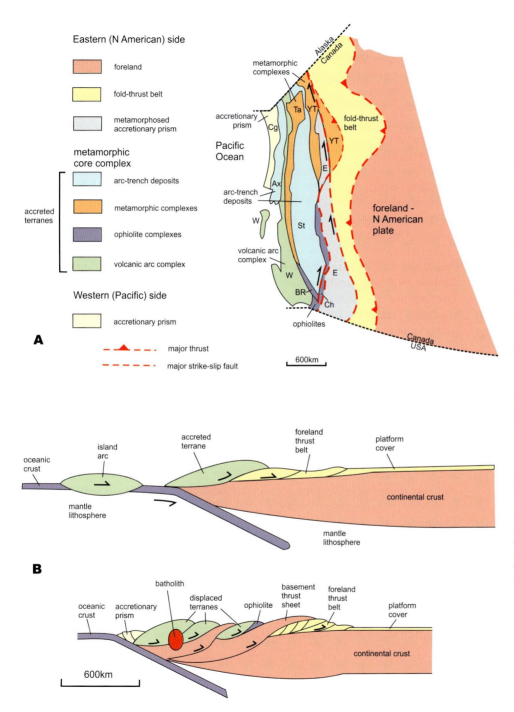

Figure 11.8 N. American Cordilleran belt: Canadian sector. **A.** Simplified map showing the main structural units of the orogenic belt. To the east of the belt is the foreland, consisting of Precambrian continental crust of the American plate with its platform sedimentary cover. The easternmost unit of the orogenic belt is the foreland fold-thrust belt; to the west of this unit is a series of thrust- or fault-bounded units, including basement slices, and various displaced terranes consisting mostly of volcanic arc material and arc–trench deposits. The westernmost unit is a late-orogenic to Recent accretionary prism. The displaced terranes have been accreted to the American plate and subsequently slid northwards along strike-slip faults. The Columbia batholith has been intruded into the western part of the orogen after terrane amalgamation. Terranes, from W to E: Cg, Chugach; W, Wrangellia; Ax, Alexander; T, Taku; TA, Tracy Arm BR, Bridge River; St, Stikine; Ch, Cache Creek;YT, Yukon-Tanana; E, Eastern assemblage. Based on the terrane map of Coney et al. (1980). **B, C.** Simplified sketch sections to show how the orogenic belt could have been formed by the amalgamation of displaced terranes; these are carried along on the oceanic plate, then scraped off the descending slab and pushed onto the leading edge of the American continental plate. This convergent movement causes the formation of (mostly) easterly-directed thrust sheets and a general thickening and shortening of the orogen. A thrust sheet with a core of metamorphosed lower-crustal material escapes to the surface, possibly as a result of gravitational spreading. Note that the terrane boundaries now are generally steep faults.

12   Ancient orogenic belts

The study and interpretation of ancient orogenic belts suffers from the disadvantage that no information is available from ocean-floor magnetic-stripe data (*see* Chapter 3 and Figures 3.5 and 3.6), and therefore, unlike more modern belts, it is not possible to track the movements of pieces of continental crust by this means. Some information about former positions of continents is available from continental palaeomagnetic data, but this only gives the palaeolatitude at a particular time, and the palaeolongitude can only be guessed at. Another problem is that, the further one goes back in geological time, the more the contemporary geological record is obscured by either cover of younger rocks, removal by erosion, or disruption by subsequent tectonic events. Nevertheless, many ancient belts have been studied in considerable detail and much is known about their history; however, their tectonic interpretation has often proved controversial, and in many cases has been subject to bewildering changes over the years!

Three comparatively well-known belts are briefly described as examples: the British sector of the Palaeozoic Caledonian belt; the Mid-Proterozoic Grenville belt of eastern Canada; and the Early Proterozoic Trans-Hudson belt of Canada. The Archaean poses particular problems of interpretation and the question of whether, or to what extent, the plate tectonic model is applicable to that Era has been much discussed.

In the final part of this chapter, this problem is addressed, taking the Superior Province of the Canadian shield and the North Atlantic craton in southern Greenland as examples.

The British sector of the Caledonian orogenic belt

The Caledonian orogenic belt, known usually as the **Caledonides**, extends from northern Norway and East Greenland, on opposite sides of the North Atlantic, through the British Isles, to Newfoundland and eastern North America, where it merges with the **Appalachian belt**. The original extent of the belt only becomes obvious when the effects of the opening of the Atlantic Ocean in the Cenozoic Era have been restored. Figure 12.1 shows a reconstruction of the relative positions of the continents during the mid-Palaeozoic, from which the Caledonian–Appalachian orogenic belt (shown in pale red) can be seen to represent a collisional orogen between **Laurentia** (North America plus Greenland), **Baltica** (northern mainland Europe) and two (possibly joined) microplates, East and West **Avalonia**. According to this reconstruction, the northern part of the British–Irish sector has resulted from collision between Baltica and Laurentia, whereas the southern part, including England and Wales, has a separate origin, being credited to

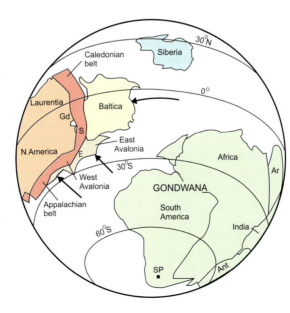

Figure 12.1 Mid-Palaeozoic reconstruction of the continents showing the Caledonian–Appalachian orogenic belt resulting from collision between Laurentia and Baltica in the northern sector, and East and West Avalonia in the southern sector. Note the positions of Scotland (S), England (E) and Greenland (Gd). Arrows show convergence direction. Based on a reconstruction by Dalziell (1997).

collision with the Avalonia microplates that have migrated from Gondwana.

The British Isles sector of the Caledonides is about 300 km wide and consists of nine distinct tectonic zones separated by major fault boundaries (Figure 12.2). These are, from north-west to south-east: North-west Foreland (Laurentia), **Moine Thrust belt**, **Northern Highlands**, **Grampian** (or **Central**) **Highlands**, **Midland Valley**, **Southern Uplands**, **Lake District**, **Welsh Basin** and **Midlands Platform**, or South-eastern Foreland (part of the East Avalonia microplate). The Northern and Grampian Highlands zones together make up the central metamorphic core of the orogenic belt. These zones were established in Scotland and England; however the Northern Highlands, Grampian Highlands, Southern Uplands and Lake District zones can also be traced across into Ireland, though with considerable differences in detail.

The Northwest Foreland

This zone consists of a mainly gneissose basement of Early Proterozoic age (the **Lewisian complex**) comparable with a formerly adjacent Early Proterozoic belt in East Greenland at the south-eastern margin of the Laurentian continent. This Lewisian basement is unconformably overlain by unmetamorphosed continental clastic red-bed sequences, the **Stoer group** and the **Torridonian Supergroup**, deposited at around 1200 Ma and 1000 Ma respectively, and by a Cambrian to Early Ordovician shallow-marine shelf sequence.

The Moine Thrust belt

This zone is a classic foreland fold-thrust belt in which the foreland sequence is involved in several major thrust packages, or nappes, each of which is divided internally by smaller thrust slices. An example of this structure is shown in Figure 12.3A, which shows several nappes containing foreland rocks overlain by a higher, exotic nappe, resting on the **Moine Thrust**, which forms the boundary with the Northern Highlands zone. The second nappe of Figure 12.3A is characterised by an **imbricate structure**, where the same Cambrian sequence is repeated many times in successive thrust slices. Structures in this zone, such as elongation lineations and deformed fossil burrows, indicate a WNW-directed shear sense.

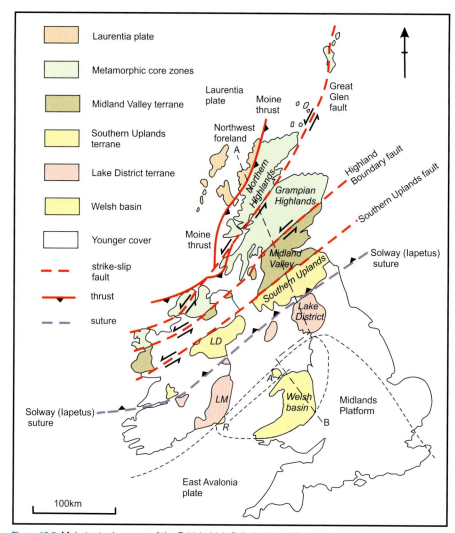

Figure 12.2 Main tectonic zones of the British–Irish Caledonides. LD, Longford Down inlier; LM, Leinster massif; A, Anglesey; R, Rosslare.

The now-exposed width of the zone is very narrow – from less than a metre to a maximum of only 10 km. However, the zone must originally have extended for a considerable distance eastwards beneath the Moine thrust and westwards to incorporate the Caledonian thrusting in the basement of the Outer Hebrides, where the platform cover has been removed by erosion, and possibly even further west onto the continental shelf.

The thrusts seem mainly to have propagated forwards, towards the foreland, such that the youngest thrusts, involving only Cambrian sediments, carry older Lewisian basement on their roofs, with the oldest, the Moine Thrust, overlying the whole package. The Moine thrust differs from the younger thrusts in being characterised by a thick band of **mylonite**, indicating derivation from considerable depth. The thrust movements are attributed to the Early Silurian **Scandian** orogenic phase, discussed below.

The Moine thrust zone is well exposed in the Assynt district of NW Scotland, which lies in the **North West Highland Geopark,** one of only eight geoparks in the UK, and internationally known as the area where Peach and Horne and their colleagues of the British Geological Survey first mapped and explained the complex thrust geometry. Much of the geology can be readily appreciated in the scenery from the roadside (e.g. Figure 12.3B) and some of the classic exposures are explained at viewpoints and at a visitor centre at **Knockan Crag,** where a section through the Moine Thrust is exposed.

The Northern Highlands

This zone contains a thick sequence of Late Proterozoic (*Neoproterozoic*) marine clastic sediments, the **Moine Supergroup**, thought probably to be the lateral equivalents of the Torridonian sequence on the foreland. They rest on a basement of Lewisian-like gneisses, and are overlain in the east by post-orogenic Devonian cover (the **Old Red Sandstone**). The Moine sequence was deposited between ~1000 Ma and ~870 Ma after the end of the **Grenville orogeny** (*see* below) and is usually regarded as resulting from the erosion of the Grenville orogen. It has been suggested that the Lewisian-like basement that underlies the Northern Highlands may actually lie within the Grenville belt and may thus have been affected by Grenvillian orogeny.

The Moine Supergroup has been intensely deformed and metamorphosed up to **amphibolite facies**, and has experienced three major orogenic events. The earliest of these is termed the **Knoydartian** and took place around 800 Ma. It is represented by syn-orogenic granitic intrusions and pegmatites and also by some metamorphic ages. The earliest recognisable foliation is associated with this metamorphic event, but the nature and extent of the Knoydartian structures have been obscured by the younger orogenic phases. Although the Moine Supergroup and its deformation have usually been considered to belong to the Caledonian orogeny, a period of more than 300 Ma separates the Knoydartian event from the main Caledonian orogenic phase, and its origin is obscure. The second phase, the **Grampian orogeny**, is the main tectonic event to

affect the Grampian zone to the south, but its effect in the Northern Highlands is much less obvious owing to the effect of the succeeding Scandian event; Grampian structures have been definitely recognised only in the **Swordly nappe**, in the eastern part of the zone.

The **Scandian orogeny** is the most important tectono-thermal event to affect the Northern Highlands and is attributed to the collision between Baltica and Laurentia in the Early Silurian. Scandian structures include ductile thrusts and related recumbent folds that are overthrust towards the west-north-west. These structures were re-folded by more upright folds with a NE–SW to NNE–SSW trend. The thrust-sense shear zones include the Moine thrust itself, the **Sgurr Beag thrust,** and the **Swordly thrust**, which lie structurally above it, to the east. Folding was accompanied by amphibolite-facies metamorphism dated at 435–420 Ma; these dates correspond to the date of the movements on the Moine thrust, at ~435–430 Ma. The structures of the Moine schists in central Ross-shire and Sutherland have attracted structural geologists since the 1950s, and many early innovative studies of superimposed folding and shear zones have been carried out there.

The Grampian (Central) Highlands

This zone contains a thick marine clastic sequence, the **Dalradian Supergroup**, consisting of sandstones, siltstones, mudstones and limestones with a total thickness of ~20 km. It is thought that this sequence was laid down on the extended passive margin of Laurentia in several **half-graben** (Figure 12.4.1; *see* also Figure 10.3). The base of the sequence is of late

ANCIENT OROGENIC BELTS

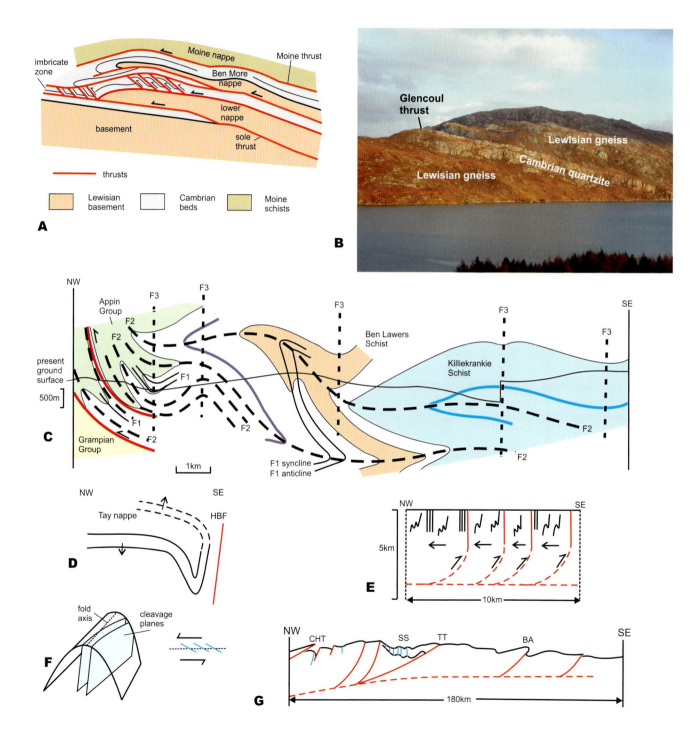

A

thrusts

| Lewisian basement | Cambrian beds | Moine schists |

imbricate zone

Moine nappe

Moine thrust

Ben More nappe

lower nappe

sole thrust

basement

B

Glencoul thrust

Lewisian gneiss

Lewisian gneiss

Cambrian quartzite

C

NW

SE

Appin Group

F3

F3

F3

F3

F3

F2

F2

F2

F1

F1

F2

F2

F2

F2

present ground surface

500m

Grampian Group

1km

F1 syncline

F1 anticline

Ben Lawers Schist

Killiekrankie Schist

D

NW

SE

Tay nappe

HBF

E

NW

SE

5km

10km

F

fold axis

cleavage planes

G

NW

SE

CHT

SS

TT

BA

180km

Figure 12.3 opposite Types of Caledonian structure. **A.** A typical foreland thrust-fold belt. Simplified cross-section across part of the Moine Thrust zone showing four superimposed nappes: lower, imbricate, Ben More and Moine nappes. The higher nappes were formed first and deformed as the younger nappes developed beneath them. The imbricate nappe consists of numerous thin slices of Cambrian strata, each sliding on a weak layer of shales that form the basal, **sole thrust** of the nappe. The slices propagate progressively forwards towards the foreland. Based on Elliott & Johnson (1980). **B.** View of the Glencoul thrust from Kylesku, Sutherland, carrying Lewisian basement over Cambrian quartzite, itself lying unconformably on Lewisian basement. **C.** Simplified cross-section through the central part of the Grampian Highlands (the metamorphic core zone) showing the effects of superimposition of F2 recumbent folds (axial surfaces shown by heavy black dashed lines) on earlier isoclinal folds (thin black lines); both are refolded by upright F3 folds (black dashed lines). The red lines are major thrust-sense shear zones. Based on Treagus (2000). **D.** The Highland boundary downbend. The anticlinal fold hinge of the recumbent Tay nappe is steepened immediately NW of the Highland Boundary Fault (HBF) to form an upright synform; only the inverted lower limb of the fold is exposed; the arrows show the direction of younging of the strata. **E.** Illustrative sketch section across the north-western part of the Southern Uplands zone showing how a series of packages of Ordovician strata, each becoming younger NW-wards, are bounded by reverse faults now steepened into near vertical attitude. The folds are typically asymmetric with a SE-up overthrust sense of movement. **F.** Sketch to show how the slaty cleavage planes in parts of the Southern Uplands are aligned clockwise from the fold axis, indicating a component of sinistral shear during compression. **G.** Simplified profile across the Welsh Basin of a notional folded surface showing the variation in fold geometry from overfolds related to thrusts to gentle folds elsewhere. The blue lines represent the attitude of the slaty cleavage in the more highly strained parts of the section in North Wales and Anglesey. CHT, Carmel Head Thrust; SS, Snowdon Syncline; TT, Tremadoc Thrust; BA, Berwyn Anticline. Based on Coward & Siddans (1979).

Precambrian age, around 650 Ma. Towards the top of the sequence are **turbidite** deposits and mafic lavas, culminating in a limestone containing Lower Cambrian trilobites. The upper part of the sequence is therefore the deeper-water equivalent to the shelf deposits of the foreland. The basement in the west is composed of gneisses belonging to the ~1,800 Ma **Rhinns complex**, but in the north-east, the Dalradian rocks lie on gneisses of the **Central Highland complex**, which are correlated with the Moine Supergroup.

The Dalradian rocks are affected by metamorphism varying from **greenschist facies** in the south-west to amphibolite facies in the north-east. The classic **Barrovian metamorphic zones** were established in the south-western to central Dalradian, but in the north-east, these are replaced by the **Buchan** type,

representing higher-temperature, lower-pressure metamorphism.

Grampian deformation has resulted in several distinct phases of folding (Figure 12.3C), usually numbered from F1 to F4. F1 consists of NW-directed ductile thrusts and recumbent folds; these are refolded by F2 overfolds that are also NW-directed in the north-west part of the zone, but in the south-east are flat lying and appear to be overthrust towards the south-east (Figures 12.3C, D). The last main phase of folding, F3, consists of more upright folds with a NE–SW trend. The main **regional metamorphism**, dated at ~475–460 Ma, accompanied the F1 and F2 folding but the F3 folds were formed during retrogressive metamorphism and were characterised by **crenulation cleavages** (*see* Chapter 7; Figures 7.1C and 7.2).

The Grampian event is attributed to a collision between the Laurentia

plate and an oceanic terrane represented now, in part, by the **Midland Valley zone**, discussed below (Figure 12.4.4). The presence of a thick ophiolite sequence in eastern Shetland, which includes a substantial upper mantle component, has prompted the suggestion that the Dalradian was overlain by a large ophiolite nappe, now removed by erosion, emplaced prior to the collision. The presence of extensive ophiolite sequences in Newfoundland and Norway is evidence that such a nappe may have had a regional extent (Figure 12.5).

The F4 folds are confined to the south-east part of the zone associated with the Highland Boundary downbend (Figure 12.3D, 12.4.6) and are attributed to late Caledonian movements linked with the collision with Avalonia (*see* below). A series of syn-orogenic plutons ranging from granites to gabbros were intruded in the period 470–460 Ma; the granites are attributed to crustal melting in the thickened orogen. The many large post-orogenic granite plutons, such as the famous Glencoe and Ben Nevis complexes (discussed in Chapter 8), are linked with a younger, Silurian, episode of subduction (Figure 12.4.6) that will be discussed below.

Major sinistral faults

The Scandian event seems to have had little or no effect in the Grampian zone, but minor tectonic activity was experienced in the region of the **Highland Boundary Fault** and **Great Glen Fault** during the late Silurian (435–420 Ma) collision of Laurentia and Avalonia. The F4 structures, including the downbend along the Highland Boundary fault, mentioned above, have

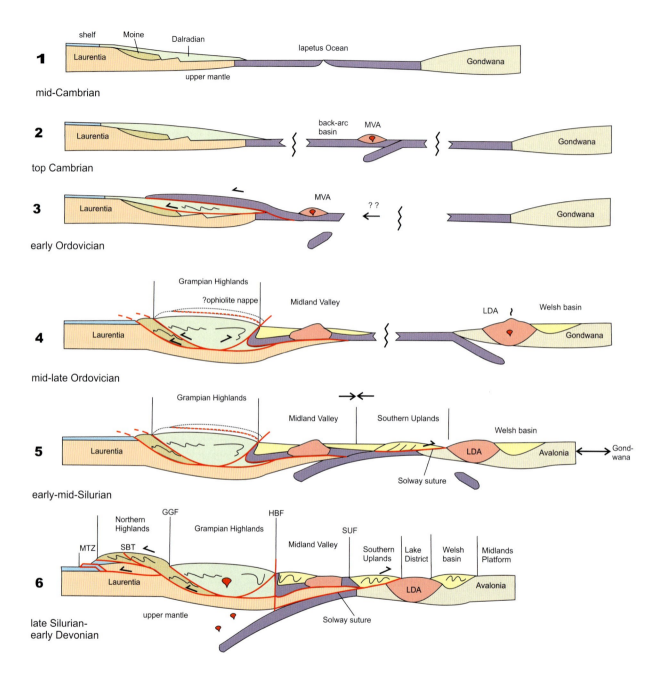

1 mid-Cambrian

shelf · Moine · Dalradian · Iapetus Ocean · Laurentia · upper mantle · Gondwana

2 top Cambrian

Laurentia · back-arc basin · MVA · Gondwana

3 early Ordovician

Laurentia · MVA · ? ? · Gondwana

4 mid-late Ordovician

Grampian Highlands · ?ophiolite nappe · Midland Valley · LDA · Welsh basin · Laurentia · Gondwana

5 early-mid-Silurian

Grampian Highlands · Midland Valley · Southern Uplands · Welsh basin · Laurentia · Solway suture · LDA · Avalonia · Gond-wana

6 late Silurian-early Devonian

Northern Highlands · GGF · HBF · Grampian Highlands · Midland Valley · SUF · MTZ · SBT · Southern Uplands · Lake District · Welsh basin · Midlands Platform · Laurentia · LDA · Avalonia · upper mantle · Solway suture

been attributed to **sinistral transpression** during this phase. The Great Glen Fault has clearly experienced a large lateral displacement, since the rocks on each side do not match up, but there has been no general agreement over the amount of displacement – estimates vary from ~160 km to over 500 km. However,

Figure 12.4 opposite Cartoon sections across the British Caledonides. An interpretation of the tectonic evolution of the crust in this sector of the Caledonides from mid-Cambrian to early Devonian times; the section is along the line A–B in Figure 12.2. **1 Mid-Cambrian (~510 Ma).** Shelf sediments and the continental-slope Dalradian sequence are deposited on stretched Laurentian basement incorporating the late Proterozoic Moine Supergroup; a wide ocean (the Iapetus Ocean) separates Laurentia from the Gondwana supercontinent. **2 Late Cambrian (~490 Ma).** The Dalradian sequence is complete; an oceanic volcanic arc, the Midland Valley arc (MVA), forms at a subduction zone within the Tethys Ocean. **3 Early Ordovician (480 Ma).** A piece of oceanic crust (from the back-arc basin of the MVA?) is obducted onto the Laurentian margin, forming the ophiolite nappe and initiating NW-directed overfolding/thrusting and metamorphism of the Dalradian in the north (the Grampian event). **4 Mid-late Ordovician (~470 Ma).** The Midland Valley arc collides with the Grampian Highlands, causing SE-directed back-folding in the south in a continuation of the Grampian event. A volcanic arc, the Lake District arc (LDA) forms at the margin of Gondwana, with a back-arc basin (the Welsh basin) behind it. **5 Early-mid Silurian (~450 Ma).** A piece of Gondwana (Avalonia) has broken away and travelled towards Laurentia. The SE-directed subduction beneath the Lake District has ceased and a new, NW-directed subduction zone formed beneath the Midland Valley as the two continental plates converge, causing an accretionary prism to form in the Southern Uplands. **6 Late Silurian–early Devonian (~400 Ma).** The two continental plates, Laurentia and Avalonia, have collided, the Solway (or Iapetus) suture marking the junction between them. The subduction zone has moved further NW, and a post-tectonic granite suite, with associated volcanics, forms in the Grampian Highlands. The Midland Valley, Southern Uplands, Lake District and Welsh Basin all experience folding and slate-grade metamorphism. Because of the oblique nature of the collision (transpressional), sinistral strike-slip faults form (including the Great Glen (GGF), Highland Boundary (HBF), and Southern Uplands (SUF) faults). The Northern Highlands experience NW-directed overfolding and thrusting (e.g. the Moine Thrust zone (MTZ) and Sgurr Beag thrust (SBT) in the Scandian event as a result of the Baltica–Laurentia collision.

which is likely, therefore, to belong to the stretched margin of Laurentia. These ophiolites, together with the more extensive oceanic assemblages in Newfoundland, East Greenland and Norway that have been thrust over the Laurentian and Baltican forelands respectively, may represent back-arc basins (e.g. *see* Figure 12.4.2). The present-day boundary of the Midland Valley terrane with the Grampian Highlands is the steeply-inclined **Highland Boundary Fault,** (*see* above and Figure 12.3D). The fact that the sequences presently juxtaposed across the Highland Boundary Fault cannot be directly matched means that it must represent a terrane boundary, and the ophiolite thus represents a suture between Laurentia and the Midland Valley terrane.

the presence of Moine sequences with similar dates on both sides of the fault suggests that it need not be regarded as a major terrane boundary.

The Midland Valley

This zone is relatively narrow – only about 80 km across – but it represents an important component of the orogen that is better represented elsewhere, as will be seen. Lower Palaeozoic rocks are only exposed in a few small inliers near the northern and southern margins of the zone, the Caledonian history of which is mostly concealed by Upper Palaeozoic cover. Nevertheless, the inliers have been intensively studied and have yielded vital information on the history of the British Caledonides. The oldest rocks are of **Arenig** age (lowermost Ordovician) and consist of an

ophiolite assemblage. The ophiolites are succeeded by a sequence of Ordovician to mid-Silurian, mainly clastic, sediments containing clasts ranging up to boulder size, derived from a volcanic arc believed to lie in the central part of the zone in the period ~490–420 Ma (Figure 12.4.2–5). The sediments are unmetamorphosed and gently folded. The Midland Valley is thus believed to represent an oceanic volcanic arc terrane that has been tectonically welded to Laurentia. Fossil assemblages both here and in the Southern Uplands have Laurentian rather than Gondwanan affinities.

Geophysical evidence indicates that the Ordovician cover lies on a crystalline basement similar in properties to that underlying the Dalradian to the north, and in Ireland, ophiolites are seen to be thrust over gneissose basement,

The Southern Uplands

This zone consists of several fault-bounded packages of steeply dipping Ordovician to Silurian strata (Figure 12.3E). The beds in each individual package become younger northwards, although the more south-eastern packages contain younger material. The individual successions in the north contain at their base Arenig-age basalts and cherts overlain by black shales that seem to have acted as weak detachment surfaces. These are believed to represent thrusts that have been steepened into their present attitude as a result of the deformation as shown in Figure 12.3E. The shales are succeeded by sedimentary sequences dominated by greywacke turbidites. The steeply inclined strata trend uniformly NE–SW and are affected by asymmetric upright folds. The shales possess a slaty cleavage. The zone is bounded in the north by the **Southern Uplands Fault**, which has experienced at least 10 km of sinistral strike-slip displacement, as there is no direct match between the Lower Palaeozoic sequences on each side. This fault is one of a set of such faults that occur throughout the belt, and it seems that some of the steepened thrusts have been re-activated in a strike-slip sense. The total sinistral offset across the zone is probably considerable, and took place during the early Devonian in the closing stages of the Caledonian orogeny. The effects of sinistral shear during this late Caledonian deformation are reflected in places by the alignment of the slaty cleavage, which is clockwise with respect to the fold axial planes (Figure 12.3F); this is consistent with the effects of sinistral **transpression** (*see* Chapter 4).

The Southern Uplands zone is bounded on its south-eastern side by the **Solway** (or **Iapetus**) **suture**, which is not exposed, but has been seismically imaged as a major discontinuity inclined at a moderate angle north-westwards beneath the zone, and which lies at a depth of around 12 km in the central part of the zone. The basement beneath the suture is interpreted as part of the Avalonian plate. This zone is represented in Ireland by the **Longford Down massif**.

The Southern Uplands has long been regarded as an accretionary prism, formed above a NW-dipping subduction zone at the north-western margin of the Avalonian microplate (Figure 12.4.5) although alternative interpretations have been suggested.

The Lake District zone

The Caledonian rocks of the English Lake District consist of a sequence of Ordovician arc-type volcanics, succeeded by Silurian marine deposits; these rest on a late Precambrian basement, which is exposed to the south in Anglesey and NW Wales. The south-western continuation of this zone in Ireland is represented by the **Leinster massif**. The zone is regarded as an Ordovician volcanic arc situated at the northern margin of the Avalonian microplate (Figure 12.4.4). These Lower Palaeozoic rocks were deformed and subjected to slate-grade metamorphism in the late Silurian, at the same time as the Southern Uplands.

The Welsh Basin

The Lower Palaeozoic rocks of Wales have been intensively studied by generations of geologists and were regarded as an example of a **eugeosyncline** in the 1930s. They comprise around 10 km of Cambrian to Silurian sediments including a large proportion of turbidites. Volcanics are an important constituent in the south-west, and especially in the north-west, in Snowdonia. The modern model is of a back-arc basin situated on thinned Avalonian crust behind the Lake District arc (Figure 12.4.4). The rocks of the zone are variably deformed in the late Silurian: tight folds with associated slaty cleavage in the north give way to more gentle folds in the south-east (Figure 12.3G).

Regional context

Figure 12.5 is an interpretation of how the zones of the British–Irish sector of the Caledonides might link up with Newfoundland to the south, and Norway and Greenland to the north. The Northern Highland and Grampian Highland zones are seen as part of a regional metamorphic 'core' zone thrust onto the Laurentian continent, and which is represented also in Newfoundland and East Greenland. Both there and in Ireland this zone is overthrust on its southeastern side by Ordovician arc terranes. South of these zones, the polarity changes to south-east-directed units, including the Ordovician accretionary prisms in Newfoundland, Scotland and Ireland, and the volcanic arc terranes on the northern margin of Avalonia. An important role is played by major strike-slip faults – the Great Glen, Highland Boundary and Southern Uplands faults being only the more obvious. The total sinistral strike-slip displacement on these is unknown, but estimates have ranged from a few hundred to over 1000 kilometres. Consequently, none of the terranes south of the Moine thrust can be directly linked to its neighbour, which makes interpretation difficult.

The strike-slip faulting results from the late Silurian collision between Laurentia and Avalonia, which must have been oblique to the plate boundary such that the convergence was partitioned into orthogonal (i.e. at right angles to the boundary) and strike-slip components. The collision between Laurentia and Baltica, which appears to have been more or less contemporaneous with the collision with Avalonia, seems to have been more nearly at right angles to the Baltica plate margin. The effect of the Baltica collision on the British Isles – the Scandian event – is not so obvious; only in the Northern Highlands have major tectonic effects been ascribed to it. It is likely that this

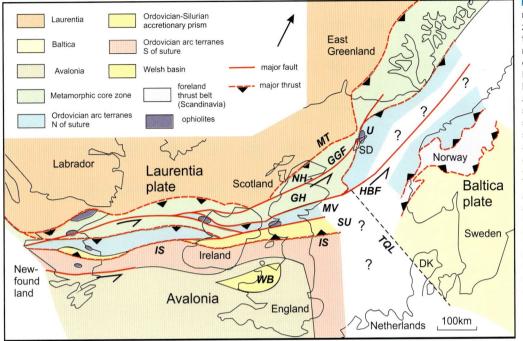

Figure 12.5 Simplified reconstruction of the tectonic zones of the Caledonides of the North Atlantic region after restoring the effect of Atlantic opening. Note the major NW-directed thrusts (red dashed lines) north of the Iapetus suture and the SE-directed thrusts south of the suture; major strike-slip faults are shown as solid red lines. The Ordovician arc terranes north of the suture are represented in Scotland only in Shetland. *MT*, Moine Thrust; *GGF*, Great Glen fault; *HBF*, Highland Boundary fault; *NH*, Northern Highlands; *GH*, Grampian Highlands; *U*, Unst ophiolite; *MV*, Midland Valley; *WB*, Welsh Basin; *IS*, Iapetus suture; *TQL*, Tornquist line; *DK*, Denmark; *SD*, Shetland. Based on a reconstruction by Dewey and Shackleton (1984).

is a result of the movements on the Great Glen Fault, which have juxtaposed terranes that were previously far to the south-west, away from the influence of the Scandian collision.

The Grenville belt

The **Grenville orogenic belt** extends from Labrador in north-eastern Canada along the south-eastern side of North America as far south as Texas, but is best preserved in Canada, where it is known as the Grenville Province. It was formed in mid-Proterozoic times during the assembly of the **Rodinia** supercontinent (Figure 12.6) as a result of collision between the early Proterozoic core of Laurentia and another continental plate or series of terranes, now removed. The Canadian sector

Figure 12.6 Possible arrangement of the continents of Laurentia, Amazonia and Baltica (part of the supercontinent of Rodinia) at ~990 Ma showing how collision with Amazonia may be responsible for the Grenville orogeny. Modified after a reconstruction by Pisarevsky *et al.* (2003).

of the belt is about 2000 km long and up to 500 km wide (Figure 12.7).

The Grenville belt has been extensively studied, and the Canadian sector in particular is comparatively well known – more so than any of the other belts of the same age. Unfortunately, however, only the north-western part of this belt is now accessible, the south-eastern part being partly obscured by the younger Appalachian–Caledonian belt. It is believed that the missing opposite side of the orogen may be represented by a Grenville-age zone

bordering the Precambrian continental block known as Amazonia, now forming the core of northern South America (Figure 12.6). However, the

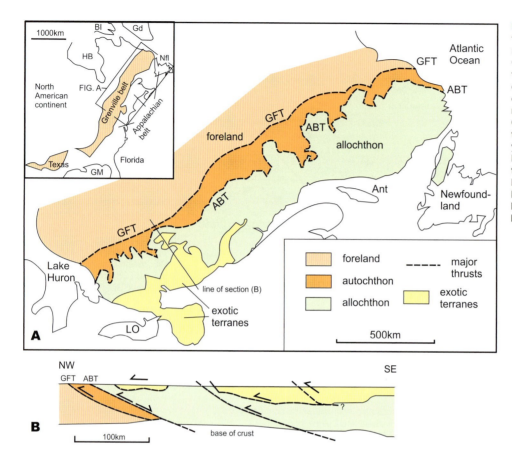

Figure 12.7 The Canadian sector of the Grenville orogenic belt. Simplified map (**A**) and cross-section (**B**) of the Canadian Grenville Province showing the outcrops of the Laurentian foreland, outer autochthon, inner allochthon and exotic terranes of the orogen 'lid'. Inset map shows the whole North American Grenville belt stretching from Labrador to Texas, bounded to the SE by the younger Appalachian belt. GFT, Grenville Front thrust; ABT, Allochthon Boundary thrust; LO, Lake Ontario; Ant, Anticosti Island; inset: Gd, Greenland; BI, Baffin Island; Nfl, Newfoundland; HB, Hudson Bay; GM, Gulf of Mexico. Based on Rivers (2009).

palaeomagnetic data for Amazonia during the period in question is not well established, and an alternative possibility is that the south-east side of the Grenville belt was formed by a series of displaced terranes bordering a subduction zone. After the end of the Grenville orogeny, a number of extensional basins were formed in the region south-east of the present Grenville belt margin, culminating in the creation of an ocean during the Cambrian period, thus moving any missing Grenville-age crust to its far side.

Structure of the Canadian sector

This part of the belt (Figure 12.7) consists of two main zones, an outer zone, bounded by a major thrust zone known as the **Grenville Front thrust** (GFT), and an inner zone bounded by a wide shear zone, named the **Allochthon Boundary thrust** (ABT). The name '**allochthon**' is applied to crust that has travelled some distance from its origin, in distinction to the '**autochthon**', which consists of locally derived crust. The foreland consists of Archaean to early Proterozoic rocks making up the core

of the Laurentia craton. These rocks are reworked but still recognisable in the outer (autochthonous) zone. The GFT descends to a depth of between 25 and 40 km, reaching the base of the crust, so that rocks originating there are now exposed at the surface. NW–SE elongation lineations on GFT shear zones indicate that thrusting was at right angles to the trend of the belt. Later extensional movements have reversed the initial thrust-sense movements on the ABT.

The Archaean and early Proterozoic rocks forming the main part of the belt are affected by high-grade metamorphism ranging from eclogite and granulite facies to upper amphibolite facies. However, the structurally uppermost units, seen in the south-west, are displaced (exotic) terranes containing mid-Proterozoic rocks that lack penetrative Grenvillian deformation and metamorphism and constitute what has been called the 'lid' of the orogen. The interior part of the belt contains a number of large syn- to post-tectonic plutons.

Tectonic evolution of the orogen

Two distinct phases of orogenic activity have been recognised. An earlier phase, known as the **Ottawan**, dated at 1090–1020Ma, was responsible for the high-grade metamorphism of the allochthon and the thrust-sense movements on the ABT; a later phase, the **Rigolet**, dated at 1090–980 Ma, was responsible for the thrusting at the Grenville Front, and the related metamorphism and deformation in the outer autochthonous zone. The evolution of the orogen has been interpreted in terms of three main phases of activity (Figure 12.8).

◆ 1. An early contractional phase between 1090 and 1020 Ma resulting in crustal thickening to around double its original thickness, resulting in the lateral extrusion of a hot middle-crustal sheet along the Allochthon Boundary thrust (ABT) (i.e. *see* the channel flow process described in Chapter 10 (*see* Figure 10.4).

◆ 2. Gravitational collapse at around 1020 Ma, resulting in the extensional movements on the ABT and the consequential cooling of the metamorphic core of the orogen.

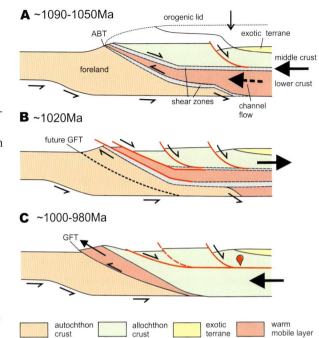

Figure 12.8 Tectonic evolution of the Grenville orogenic belt. Simplified sketch sections illustrating the tectonic evolution in three stages. 1, ~1090–1050 Ma: tectonic over-thickening of the crust causes the warmed lower mobile crust to extrude along the Allochthon Boundary thrust (ABT); this mobile sheet is bounded by two shear zones with opposed senses of movement. 2, ~1020 Ma: a pause in tectonic compression allows the orogen to partially collapse and cool. 3, ~1000–980 Ma: renewed compression causes activity to migrate towards the foreland, causing the warm lower crustal layer to move up along the Grenville Front thrust (GFT). Based on Rivers (2009).

◆ 3. Renewed contraction spreads outwards into the foreland causing the uplift of the basement along the Grenville Front thrust (GFT) at around 1000–980 Ma.

The Grenville orogen forms an interesting contrast with both the Alpine and the Caledonian orogens described above, in that it was comparatively long-lived (nearly 100 Ma) and produced high temperature conditions over a wide area. Because erosion has revealed deeper crustal levels than is the case in these younger belts, it is easier to investigate the deep structure and to reconstruct the tectonic processes involved.

The Trans-Hudson orogen: the making of Laurentia

The Early Proterozoic (*Palaeo-proterozoic*) **Trans-Hudson orogenic belt**, as the name suggests, is centred on Hudson Bay and resulted from the amalgamation of a number of older cratons and terranes of Archaean age during the period ~1920 Ma to ~1800 Ma; a process involving a series of separate collisions ending with the final collisional event that brought the Trans-Hudson orogeny to a close. The end result was the formation of the Precambrian core of North America (Figure 12.9) which extends from Wyoming in the northern USA to Baffin Island in northern Canada, a distance of about 3500 km. In the west it is obscured by younger platform cover, but the exposed width from there to

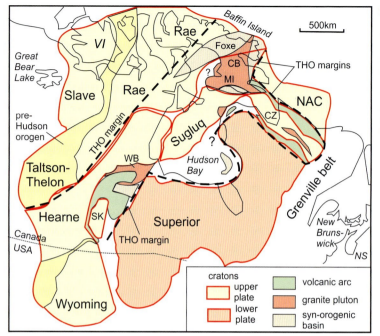

Figure 12.9 The Trans-Hudson orogen. Simplified sketch map of northern North America showing the various components of Laurentia. The various cratons and terranes making up the upper plate of the main Hudsonian collision orogen are shown in yellow; the lower plate, the Superior craton, is shown in pale orange. Important areas of oceanic volcanics and granitic plutons are also shown. *NAC*, North Atlantic craton; *CZ*, Core Zone terrane; *MI*, Meta Incognita terrane; *CB*, Cumberland batholith; *WB*, Wathaman batholith; *SK*, Sask terrane; *CS*, Cape Smith belt; *VI*, Victoria Island; *NS*, Nova Scotia; *TB*, Torngat Belt; the outline of Greenland has been omitted for clarity. Based on Corrigan *et al.* (2009).

Labrador in the east, where it is cut off by the younger Grenville belt, measures over 3000 km. Together with the greater part of Greenland, this Precambrian continent is known as **Laurentia**.

The Trans-Hudson belt itself varies in width from about 500 km in the southwest to over 1300 km in the Hudson Bay embayment. It is bounded on its northwest side by the Archaean **Slave** craton together with that part of the Archaean **Rae craton** unaffected by Hudsonian overprint. Between these two Archaean cratons is another Early Proterozoic orogenic belt, known as the **Taltson-Thelon belt**, formed before the Trans-Hudson orogen. North of the Superior craton, a branch of the Trans-Hudson belt turns southeast, bounded by the Archaean **North Atlantic craton** on its northeast side; here the belt is narrower – about 400 km wide. North of Baffin Island,

the main belt, known here as the **Foxe belt**, continues into west Greenland.

As is the case with the Grenville belt described above, this ancient orogen has completely stabilised, much of the younger cover has been removed, and relatively deep-level rocks are now exposed at the surface. Consequently, with the aid of a good coverage of high-quality age dates, it is possible to determine the position of the sutures between pieces of crust of different age, and thus to piece together the possible sequence of events leading up to the final amalgamation of the Laurentian continent. Knowledge of the palaeomagnetic record of the Archaean components of the orogen is sparse, and therefore the relative position of the component cratons prior to amalgamation is mostly guesswork, although the directions of

convergence can often be deduced by structural evidence from thrusts, etc.

Structure of the Trans-Hudson orogen (THO)

The western part of the THO (Figure 12.9) is known as the **Western Churchill Province** and consists of the **Hearne craton** and the **Sugluq craton**, together with the eastern Rae craton and the zones of collision between them. In the north, the collision was between the Rae and **Meta Incognita** cratons, resulting in the **Foxe orogenic belt**. This collisional belt also affects the sedimentary cover of the Rae craton (the **Piling Group**) consisting of a shallow-marine shelf sequence overlain by a foredeep basin – the **Penrhyn–Piling basin**.

The eastern part of the THO is usually known as the **Reindeer zone**. This zone contains a series of oceanic volcanic arcs, back-arc basins and oceanic plateaux (**La Ronge–Lynn Lake**, **Flin-Flon–Glennie**, and **Parent–Spartan** oceanic arc terranes) originating in the former oceanic area between the Hearne and Sugluq cratons on the western side and the Superior craton on the eastern side, and which became accreted to the Hearne and Sugluq cratons before the final closure of the ocean.

The north-eastern arm of the THO, termed the **Eastern Churchill Province**, consists of two separate collisional zones between the North Atlantic and Superior Archaean cratons – the **Torngat belt** on the north-east side and the **New Quebec zone** on the south-west side, separated by a 200 km-wide central zone of reworked Archaean basement known as the **Core zone**. Both the Torngat and New Quebec zones contain deposits ranging from shelf to oceanic assemblages strongly deformed and thrust towards the south-west. The western extension of the Core zone in Baffin Island, the **Meta Incognita craton**, is separated from the Sugluq craton by the western extension of the New Quebec subduction zone. Much of the upper plate here is obscured by the large **Cumberland batholith**. The continuation of the New Quebec zone around the northern margin of the Superior craton consists of a fold-thrust belt known as the **Cape Smith belt**, which contains a 5 km-thick ophiolite complex (the **Purtuniq ophiolite**) – one of the earliest to be recognised.

A series of continental volcanic arcs, now represented mainly by granitic plutons such as the Wathaman and Cumberland batholiths, line the outer margins of the Hearne, Sugluq, North Atlantic, and Core zone cratons, indicating that subduction zones formerly bordered these continental terranes, which therefore represent the upper plate of the THO collisional belt.

Tectonic evolution of the THO

Four main phases of the orogeny have been recognised, each representing a series of events culminating in collisions, moving progressively from west to east across the THO (Figure 12.10). These are as follows.

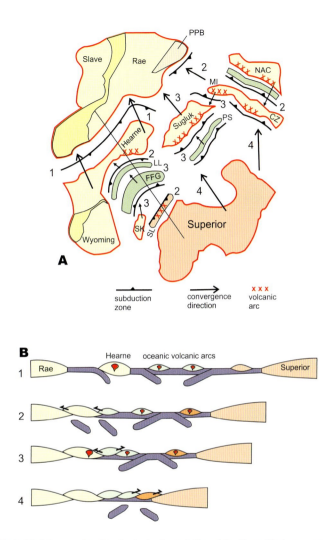

Figure 12.10 A. Sketch map showing the tectonic evolution of the Trans-Hudson orogen in terms of four stages: **1**, 1920–1890 Ma: the Hearne–Wyoming craton converges with the combined Slave–Rae craton; volcanic arcs develop in the ocean between the Hearne and Superior cratons. **2**, 1880–1865 Ma: the La Ronge–Lynn Lake volcanic arc accretes with the Hearne craton; the NAC collides with the Core Zone terrane, and the Meta Incognita craton with the Rae craton, with consequential closure and deformation of the Penrhyn–Piling basin; subduction beneath the Superior craton forming the Snow Lake volcanic arc. **3**, 1865–1840 Ma: the Flin-Flon–Glennie and Parent–Spartan volcanic arcs converge respectively with the Hearne and Sugluq cratons; the Meta Incognita terrane converges with the Sugluq terrane and the Sask terrane with the Flin-Flon–Glennie arc; granitic plutons emplaced along the SW margins of the Hearne, Sugluq and Meta Incognita cratons. **4**, 1830–1800Ma: the main, end-Hudsonian event: the Superior craton collides with the combined upper-plate cratons already amalgamated. NAC, North Atlantic craton; MI, Meta Incognita; CZ, Core Zone; LL, La Ronge–Lynn Lake arc; FFG, Flin-Flon–Glennie arc; PS, Parent–Spartan arc; PPB, Penrhyn–Piling basin; SL, Snow Lake belt. Based on reconstruction by Corrigan et al. (2009). **B.** Cartoon sections illustrating the possible tectonic evolution of the Trans-Hudson orogen along the line of section shown on Figure B. Stage 1: 1920–1890 Ma; 2: 1880–1865 Ma; 3: 1865–1840 Ma; 4: 1830–1800 Ma.

◆ 1. The **Snowbird phase** (~1920–1890 Ma): convergence and collision of the Hearne and Rae cratons together with the deformation of the intervening volcanic-arc deposits; development of the oceanic volcanic arc terranes in the ocean between the Hearne, Sugluq and Superior cratons.

◆ 2. The **Reindeer–Foxe phase** (~1880–1865 Ma): accretion of the La Ronge–Lynn Lake oceanic arc terrane to the Hearne craton; collision of the Meta Incognita craton with the Rae craton; and consequent closure and deformation of the Penrhyn–Piling basin. In the Eastern Churchill Province, collision between the NAC and the Core zone to form the Torngat belt.

◆ 3. The **Wathaman phase** (~1865–1840 Ma): accretion of the Parent–Spartan and Flin-Flon–Glennie volcanic arc terranes to the Sugluq and Hearne cratons; collision of the Sask terrane with the Flin-Flon–Glennie volcanic arc; emplacement of granite batholiths, incuding the Wathaman and Cumberland batholiths, along the south-west margins of the Hearne, Sugluq and Meta Incognita cratons; formation of a magmatic arc (the **Snow Lake belt**) at the north-western margin of the Superior craton.

◆ 4. Main **Hudsonian orogeny** (~1840–1800 Ma): emplacement of continental-arc plutons along the margin of the Meta Incognita craton and in the Core zone; (from ~1830 Ma) terminal collision of the Superior craton with the previously accreted collage (the Hearne, Rae, Sugluq and NAC–Core zone cratons plus the intervening

volcanic arc terranes, accompanied by a widespread tectono-thermal overprint. Collision started in the south and migrated around the Superior craton, ending in the New Quebec belt.

The Trans-Hudson orogen is one of the earliest examples of an orogenic belt where 'modern' plate tectonic processes can plausibly be adduced to explain the overall structure and sequence of tectonic events. This is due to the fact that separate terranes and their intervening sutures, together with subduction-related magmatism, can be identified. The preservation of identifiable shelf sequences, such as the Piling Group, on bordering cratons, well-developed ophiolites (the Purtuniq ophiolite), foreland thrust belts (e.g. along the Superior craton margin) and foreland basins (e.g. the Penrhyn–Piling basin) all point to the operation of plate tectonic processes similar to those of later periods. The same cannot be said of the Archaean, which we shall now discuss.

The Archaean
The **Archaean Eon** extends from 4000 Ma ago until the beginning of the Proterozoic at 2500 Ma, a time span three times as long as the Phanerozoic. However, evidence of tectonic processes becomes increasingly difficult to establish, the further back in time we seek it. Rocks of Archaean age make up a large proportion of the Precambrian cores of all the continents, but in many cases have been intensively reworked in subsequent orogenies, as can be seen, for example, in the Archaean terranes incorporated into the Early Proterozoic Trans-Hudson orogen just discussed. Most areas of Archaean outcrop (i.e. the

Archaean cratons) are therefore fragments of larger areas that have been disrupted by younger events, and it is rare to find evidence of boundaries between Archaean areas that have demonstrably experienced orogenesis and adjoining areas that might represent a contemporary stable foreland. Most Archaean cratons were formed in the period 2700–2600 Ma and are thought by some to represent pieces of a late Archaean supercontinent. Evidence of earlier Archaean continental cratons of any size is sparse. However, Archaean rocks of the **Pilbara craton** in NW Australia and the **Kaapvaal craton** in South Africa had become stabilised at 2800 Ma and 2900 Ma respectively, and could be said to represent the continental forelands of younger Archaean orogenic belts.

For many years there was considerable debate about whether or not 'modern' plate-tectonic processes operated in the Archaean, but this debate has been largely resolved. Increasing knowledge of the detailed composition of several Archaean cratons has indicated that they can be explained by the processes of accretion of continental, oceanic and volcanic arc material in much the same way as more recent orogenic belts. Moreover, the composition of the igneous rocks is similar, though not identical, to the magmatic products of modern subduction zones. Differences include a preponderance of **greywackes** in the sedimentary sequences compared to typical platform sequences, a lack of high-pressure **blue schists** and Phanerozoic-type ophiolite assemblages, a more **tonalitic** composition of the granitic plutons, and the prevalence of **komatiites** (ultrabasic volcanics). Some of these differences

have been ascribed to more vigorous oceanic convection and shallower subduction. It seems highly probable that the heat flow through the Earth's surface was considerably higher during the Archaean than at present, which helps to explain these observations.

In describing examples of orogenesis in the Archaean, therefore, we look at two regions where the geology is comparatively well known, with comprehensive dating coverage, but whose limits cannot be established because of the fragmentation referred to above, so that the former extent of these Archaean orogenic belts is uncertain. Archaean regions are commonly divided into two basic types – **granite–greenstone** and **high-grade gneiss**, but these distinctions are based mainly on the grade of metamorphism and degree of deformation at the current level of exposure, and do not reveal any fundamental distinction. The **Superior Province** of Canada contains typical examples of granite–greenstone terrain, and the North Atlantic craton of southern Greenland (*see* Figure 12.9) is often cited as an example of a high-grade gneiss terrain.

The Superior Province of Canada

This large Archaean outcrop (*see* Figure 12.9) forms the core of the Precambrian shield of North America around which are draped the Proterozoic belts of the Trans-Hudson and Grenville orogens. It consists of a number of oceanic and continental terranes, formed over a period of ~3700 Ma to ~2650 Ma, that have accreted along a convergent plate margin about 1000 kilometres long in a series of separate collision events.

There are eleven major zones, known locally as sub-provinces or '**domains**',

each characterised by a distinctive rock suite or metamorphic grade. They range in width from 40 km to 200 km and are generally separated by a fault or boundary zone across which the metamorphic grade changes abruptly. The province as a whole is dominated by granitic rocks. In some cases, rock units can be correlated across the domain boundaries so that the domains do not all represent separate terranes.

There are three major types of domain: 1) the *high-grade gneiss* type consisting of **orthogneisses** and granitoid rocks with only minor **supracrustal** components; 2) the *granite-greenstone* type; and 3) the *metasedimentary gneiss* type. Some granitic domains, such as Berens River and Winnipeg River, contain components over 3000 Ma in age and may represent micro-continents about which the younger supracrustal domains were

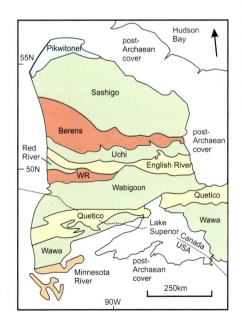

amalgamated. Metamorphism in the granite–greenstone domains ranges from greenschist to amphibolite facies, and in the metasedimentary domains is typically amphibolite facies. The domains are believed to have been tectonically amalgamated into a coherent craton at about 2700 Ma ago.

The narrow **Pikwitonei domain** at the north-western margin of the Superior Province (Figure 12.11) is a typical high-grade gneiss belt with granulite-facies metamorphic assemblages in a variety of lithologies, including 3400–3000 Ma orthogneisses and more massive younger tonalitic plutons. Several phases of high-grade metamorphism occurred in the period 2716–2642 Ma, the youngest being associated with development of the major NE–SW-trending deformation.

The **Sashigo** and **Berens River** domains, which together form a block over 300 km wide, contain isolated, thin **greenstone belts** surrounded by extensive granite and gneiss units, the Berens domain being almost entirely dominated by granitic rocks. Granitic basement over 3000 Ma in age has been identified beneath the greenstone cover. The Pikwitonei

Figure 12.11 The Superior Province. Simplified map of the western Superior Province showing the subdivision into zones (known in Canada as domains) based on their predominant rock type. Four main types are recognised: 1) high-grade gneiss domains (blue); 2) granite–greenstone domains (green); 3) granite domains (pale red) and 4) metasedimentary domains (yellow). Several (but not all) domains represent separate terranes. The Wawa is a 'super-terrane' (see text). WR, Winnipeg River. Based on Card & Cieselski (1986).

domain probably represents the high-grade equivalent of the Sashigo granite–greenstone assemblage.

The narrow **Uchi domain** consists of a series of thin, tightly-folded, E–W-trending belts of metavolcanic and metasedimentary rocks intruded by elongate granodioritic to tonalitic intrusions.

The **English River domain** is an E–W linear belt up to 60 km wide containing high-grade, 2700–2650 Ma-old metasedimentary, metavolcanic and granitic gneisses. The metasediments include metagreywackes believed to result from the erosion of the neighbouring granite–greenstone domains.

The **Winnipeg River domain** is a granite–greenstone belt similar to the Uchi but dominated by granitoid rocks. Highly metamorphosed **supracrustal** rocks (i.e. sediments and volcanics) are intruded by granodioritic to dioritic plutons emplaced between 2710 and 2660 Ma. 3319 Ma-old tonalitic gneisses are also present.

The **Quetico domain** is a metasedimentary belt consisting predominantly of 2700–2690 Ma-old metagreywackes believed to be derived from the adjoining volcanic domains.

The **Wawa domain** is the western extension of the **Abitibi** granite–greenstone domain of the eastern Superior Province. It is regarded as a '*superterrane*' and consists of a subduction-accretion complex of several separate intra-oceanic and volcanic island-arc belts formed during the period 2720–2680 Ma and amalgamated in a collisional event with the adjoining terranes to the north and south at 2690–2670 Ma.

The **Minnesota River domain** is located in the extreme south-western corner of the Superior Province and lies wholly within the USA. It consists of ~2700 Ma metavolcanic and metasedimentary rocks on a basement of granitoid gneisses up to 3600 Ma old – the oldest rocks presently known within the Province.

Although the Superior Province has been thought of as a ~2700 Ma collisional orogen, it is now bounded by younger orogenic belts on both sides, and neither the nature of the original bordering continental plates, if they existed, nor the original dimensions of the orogen, can now be established. It may be more realistic to visualise the orogen as a collection of accretionary terranes. However, the existence of granitoid gneisses ranging in age from 3600 to 3000 Ma indicates that several distinct continental fragments have been caught up in the orogen and may represent pieces of pre-existing continents.

The North Atlantic craton (NAC)

This region, often taken as the 'type example' of a high-grade gneiss terrain, lies partly in south Greenland and partly in eastern Labrador, and forms the north-eastern foreland of the Trans-Hudson orogen described above (*see* Figure 12.9). It is much smaller than the Superior craton to the south – only about 500 km at its widest on the western coast of Greenland, and much of it is obscured by the Greenland icecap. However, the exceptional coastal exposures have been intensively studied. The craton reveals a long history of geological activity spanning around 1500 million years that includes at least three major tectonic events that could be analogous to more modern orogenies; only the last of these might be considered to be a craton-wide orogeny. Because of the length of time represented by the Archaean, it has been found convenient to subdivide it into early, middle and late periods – **Eo-archaean** (4000–3500 Ma), **Meso-archaean** (3500–3000 Ma) and **Neo-archaean** (3000–2500 Ma) respectively.

Granitoid rocks, generally granitic to tonalitic in composition, are the dominant rock type and are typically deformed into gneisses. These contain elongate sheets and lenses from around one metre to over a kilometre in width, of supracrustal rocks that are often strongly folded. The supracrustal units include basic to ultrabasic igneous rocks, including both volcanics and intrusives, which are comparable geochemically both with the greenstone belts of the Superior Province and with modern oceanic crustal sequences. The supracrustal units also include mica-schists and marbles, whose association with the oceanic volcanics suggests that they were probably laid down on oceanic crust.

According to the chronology established in the early mapping of the western NAC, the earliest rocks in the NAC, a group of supracrustal volcanic and sedimentary rocks of Eo-archaean age (~3900 Ma) known as the **Isua supracrustals**, were invaded by granitic gneisses (the **Amitsoq gneisses**). After the resulting Eo-archaean complex had been formed, a second set of supracrustal rocks, of Meso-archaean age (the **Malene supracrustals**), were deposited, and involved in a major deformation that resulted in the tectonic interleaving of Amitsoq and Malene units at ~3000 Ma ago. These in turn were invaded by further granitic intrusions, now gneissose, (the **Nuuk gneisses**) and subjected to widespread deformation and high-grade metamorphism at ~2800 Ma ago. However, more recent, detailed mapping and precise dating have caused this sequence to be revised and a number of separate terranes identified, each of which is considered to have had a different history to its neighbour, prior to their amalgamation in the late Archaean.

Figure 12.9 shows a map and cross-section of part of the gneiss complex north of the town of Nuuk (formerly Godthaab), which gives an impression of the complexity of the area. The Neo-archaean deformation has resulted in the formation of a layered complex, aligned roughly N–S, where terranes of varying age have become tectonically amalgamated by a series of gently inclined shear zones and subsequently refolded by upright folds. Six separate terranes have been recognised in this area. The terrane boundaries are cut by a late-Archaean granite. Several of these terranes contain folded and disrupted supracrustal layers, one of which, the famous **Isua** greenstone belt, is highlighted in the north of the map. Two generations of (high-grade) greenstone belts have been recognised, each containing accreted units of oceanic and volcanic arc material similar to those described in the Superior Province.

It is difficult to compare such an area with more modern orogenic belts, particularly as the scale is so small, compared even with the Superior Province. The area of Figure 12.12 is only about 50 km across and the individual terranes recorded there are of the order of 10–20 km wide. Despite the fact that oceanic and continental crustal units have been accreted together, the individual terranes are more like thrust slices than separate micro-continents, and some have been amalgamated before the late Archaean orogenic event. In terms of structural and metamorphic state, perhaps the best analogy would be with the mid- to lower-crustal sections of the Himalayan or Caledonian orogenic belts described earlier, where strong ductile deformation at elevated temperature has produced similar results.

Archaean orogeny

The recognition in both the Superior Province and the North Atlantic craton of igneous assemblages comparable to present-day oceanic plateaux and volcanic island arcs, which have

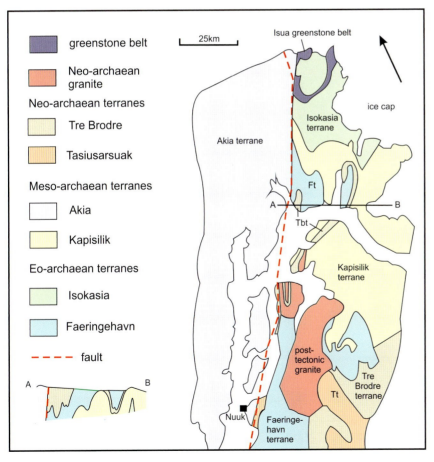

Figure 12.12 Archaean terranes of the Nuuk area. Map and E–W section across the area north of Nuuk (formerly Godthaab) in southern west Greenland, showing the various terranes that have been recognised in the area. Ft, Faeringehavn terrane; TBt, Tre Brodre terrane; Tt, Tasiusarsuak terrane; Based on Friend & Nutman (2005).

been brought together by a process of accretion, lends strong support to the proposition that plate tectonic processes similar to today's operated in much the same way as far back as ~4000 Ma ago when the oldest rocks we can now see were formed.

The contrasts between the Superior and North Atlantic cratons may be due mainly to the difference in level of exposure, as the NAC forms the upper plate of the collision with the Superior Province, thus exposing its lower crust, whereas the latter is surrounded by continental plates that have been thrust over it. The large granitic plutons that are such a characteristic feature of the granite–greenstone terrains, for example, may be represented in the high-grade lower crust by their much less prominent roots.

Although differences in detail certainly exist between Archaean and post-Archaean orogenic belts, these can probably be explained to some extent by variations in the plate tectonic mechanism arising from the greater Archaean heat flow. It is difficult to obtain enough information about the extent or nature of orogenic belts or orogenic processes further back than the Neo-archaean because of the disruption of the pre-existing crust by the pervasive late-Archaean orogenies.

Appendix

Geological Time (Table A1)

Geological time is divided into **Eons** – from oldest to youngest: **Archaean**, **Proterozoic** and **Phanerozoic**; the Phanerozoic is subdivided into **Eras** – **Palaeozoic** to **Cenozoic**; and further into Periods (**Cambrian** to **Quaternary**). Dates are given in million years before present. B. The geological column drawn to scale; note that the Precambrian represents nearly 9/10ths of geological time.

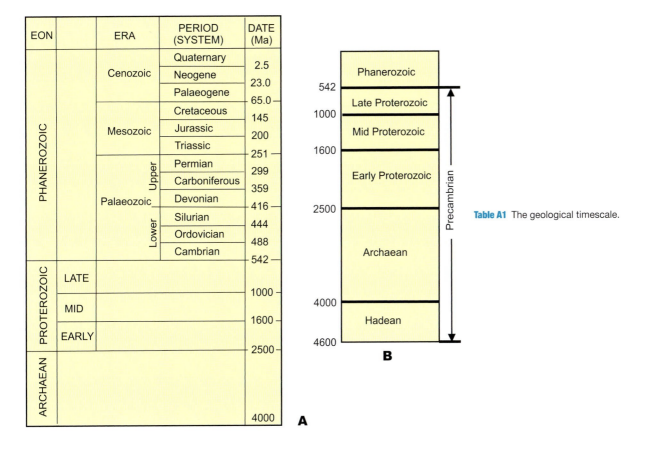

Table A1 The geological timescale.

Classification of Igneous Rocks (Table A2)

Igneous rocks are subdivided according to 1) grain size (coarse to fine); and 2) silica content (**acid** to **ultrabasic**). The main types of igneous rock are distinguished by grain size: coarse (average > 5 mm), medium (1–5 mm), fine (< 1 mm); and by silica content: acid (characterised by excess silica, giving quartz; intermediate (characterised by abundant feldspar plus silica-rich ferromagnesian minerals hornblende and biotite); basic (feldspar plus silica-poor ferromagnesian minerals pyroxene and olivine) and ultrabasic (little or no feldspar, plus silica-poor olivine and pyroxene). Igneous rocks may be further sub-divided according to their main constituent mineral(s); e.g. a granite may be further specified thus: biotite-granite or hornblende-granite, etc. Granites are further subdivided according to the ratio of alkali-feldspar to plagioclase, thus: **alkali granite**: alkali-feldspar > plagioclase; **granodiorite**: plagioclase > alkali feldspar; **tonalite**: alkali-feldspar absent.

Table A2 Main types of igneous rock.

grain size	acid	intermediate	basic	ultrabasic
coarse	GRANITE	DIORITE	GABBRO	PERIDOTITE
medium	MICRO-GRANITE	MICRO-DIORITE	DOLERITE	
fine	RHYOLITE	ANDESITE	BASALT	KOMATIITE
main minerals	feldspar quartz mica hornblende	feldspar hornblende biotite	feldspar pyroxene olivine	pyroxene olivine

Classification of Sedimentary Rocks (Table A3)

Sedimentary rocks are subdivided firstly according to origin, whether they are composed of: 1) fragments of older rocks (**clastic**) or 2) of chemical or organic origin (e.g. limestone or coal); clastic rocks are further divided according to grain size (coarse to fine); clastic rocks composed of mixed grain size and composition are termed **greywacke**; some rocks may be chemical or organic and clastic, e.g. limestone.

clastic	coarse	conglomerate, breccia
	medium	sandstone
	fine	siltstone
	very fine	mudstone, shale
mixed clastic	greywacke	
chemical/ organic	limestone gypsum rock salt	coal ironstone limestone

Table A3 Some common types of sedimentary rock.

Classification of Metamorphic Rocks (Table A4)

Metamorphic rocks are subdivided according to: 1) the type of sedimentary rock from which they are derived, i.e. mudstone, siltstone or sandstone become respectively **pelite**, **semipelite**, **semi-psammite**, or **psammite**, and limestone becomes **marble**; and 2) in the case of the fine-grained clastic rocks, further subdivided according to metamorphic grain size: from fine to coarse into **slate**, **phyllite**, **schist** and **gneiss**. There are no separate metamorphic names for the coarse-grained equivalents of coarse clastic rocks or marbles. Metamorphosed pelitic rocks usually possess a planar fabric produced by the parallel alignment of tabular minerals such as mica or hornblende, and the metamorphic rock may be further characterised by adding the name of the appropriate mineral, thus: biotite-schist, hornblende-gneiss. Gneisses are also formed by deformation of igneous rocks, and are distinguished by a banded or lensoid structure formed by the segregation of the light-coloured minerals such as quartz and feldspar from the darker, such as mica or hornblende.

	fine ⟶ coarse		carbonate
sedimentary rock	mudstone, shale siltstone	sandstone	limestone, dolomite
metamorphic equivalent	pelite semipelite	semi-psammite psammite	marble
fine ↓ coarse	slate phyllite schist gneiss	no metamorphic term for coarse varieties	

Table A4 Some common types of metamorphic rock.

Glossary

Abitibi domain [111]: tectonic zone in the **Superior Province** consisting of a **granite–greenstone terrain** interpreted as a **subduction–accretion** complex of intra-oceanic and volcanic island-arc belts amalgamated in a ~2700 Ma collisional event with the adjoining terranes (see Figure 12.11).

accretionary prism (or accretionary wedge) [82]: an accumulation of **clastic** sediments and volcanic debris occupying the **trench** and **continental slope** of an **active continental margin**, which is piled up in a series of folded and thrust slices (Figure 10.2).

acid (igneous rock) [116]: containing **quartz,** formed from a magma that is oversaturated in **silica** (see Table A2).

active bending [64]: **folding** caused by **faulting**, induced by the upward or downward movement of a layer under compression (see Figure 6.10A).

active continental margin [80]: corresponding to a **subduction** zone or **transform fault**.

African rift (system) [15-16]: a zone of rift valleys extending from the Red Sea coast in Ethiopia in the north to the Indian Ocean coast in Mozambique in the south, characterised by vulcanicity and extensional fault movements (see Figure 3.9B).

Alexander terrane [95-96]: displaced **terrane**, part of the **Cordilleran orogenic belt** of North America, which, together with the **Wrangellia terrane**, joined the American **plate** in the Cretaceous (see Figure 11.8).

Allochthon Boundary thrust [106]: major **ductile shear zone** marking the NW boundary of the inner allochthonous (far-travelled) units of the **Grenville orogenic belt** (see Figure 12.7).

Alpine orogeny [91-95]: series of orogenic episodes affecting southern Europe, resulting from the collision of Eurasia and Africa during the **Cenozoic Era**, and culminating in the **Miocene** (see Figure 11.5).

Altyn Tagh fault [41]: major sinistral strike-slip fault in the Asian plate between the Tibetan plateau and the Tarim basin, thought to accommodate some of the northwards movement of the Indian plate (see Figure 11.2).

Amazonia [105]: **Mid-Proterozoic** continent comprising much of present-day South America, thought to have collided with **Laurentia** during the **Grenville orogeny** (see Figure 12.6).

Amitsoq gneisses [112]: group of broadly granitic **gneisses** in the West Greenland sector of the **North Atlantic craton.**

amphibole: member of a group of ferromagnesian **silicate** minerals with complex compositions; **hornblende** is the commonest variety.

amphibolite facies [99]: 'medium-grade' metamorphic assemblage characterised by high temperatures and moderate pressures.

andesite [116]: fine-grained **intermediate igneous** rock (see Table A2).

anhydrite [76]: magnesium sulphate; a common component of **evaporites.**

anticline [48]: a **fold** where the older beds are in the centre or **core** (see Figure 6.3).

antiform [48]: a **fold** that closes upwards (see Figure 6.4A).

antithetic shears [62]: a set of **shear** planes making a large angle with the main **foliation** and with the opposite sense of shear (see Figure 7.4C).

Appalachian (belt) [97]: **orogenic** belt along the south-eastern margin of North America; the northern part of which, from New York to Newfoundland, resulted from the **Caledonian orogeny** (see Figure 12.1).

Apulia (microplate) [92]: micro-continent originally belonging to the African plate but which became separated from it during the Cretaceous; it now forms the southern **foreland** to the **Alpine orogenic belt** in the Adriatic region (see Figure 11.5).

Archaean Eon [115]: unit of geological time spanning the period 4000–2500 Ma (see Table A1).

Arenig [103]: lowermost unit of the **Ordovician System** (see Table A1).

aseismic [9]: lacking in earthquake activity.

asthenosphere [8]: weak layer beneath the **lithosphere**, capable of solid-state flow, over which the **plates** move (see Figure 2.4).

asymmetric folds [48]: where the **fold limbs** are of unequal length (see Figure 6.4F).

augen [61]: lens-shaped body (e.g. a deformed large crystal) within a **schist** or **gneiss.**

autochthon [106]: displaced part of an orogenic belt identified as belonging to the adjacent continental plate.

Avalonia [97]: a small continental micro-**plate** (or plates) consisting of parts of eastern North America and western Europe (including England), that was detached from **Gondwana** during the early Palaeozoic and collided with **Laurentia** during the **Caledonian orogeny** (see Figure 12.1).

back-arc basin [17]: sedimentary basin situated on the upper plate of a

subduction zone behind the volcanic arc (see Figure 3.11).

back-arc spreading [17]: process of forming **oceanic crust** behind a volcanic arc by extension of the upper plate above a **subduction zone.**

back-thrusting [94]: thrusting directed in the opposite direction to the main **thrust** movements.

Baltica [97]: a continent consisting of most of northern Europe, which existed during the **Lower Palaeozoic**, prior to the **Caledonian orogeny** (see Figure 12.1).

Barrovian (metamorphic zones) [101]: a sequence of metamorphic assemblages formed under conditions of moderate temperature and pressure increases.

basic (igneous rock) [116]: without **quartz**, formed from a magma that is undersaturated with **silica**; the ferromagnesian minerals are typically **pyroxene** and **olivine** (see Table A2).

basin [53]: a structure whose surface **dips** radially inwards, e.g. a **synclinal pericline**; a sedimentary basin is any depressed area receiving or containing sediments.

batholith [69]: a **pluton** with a very large horizontal extent and great thickness, often with no determinable floor.

bayonet structure [68]: a feature at the margin of a **dyke**, shaped like a bayonet, caused by the breaking of a bridge of host rock between the ends of two overlapping intrusions (see Figure 8.2E).

Berens River domain [111]: tectonic zone within the **Superior Province**: a **granite–greenstone terrain** almost entirely dominated by granitic rocks (see Figure 12.11).

Betic Alps [92]: branch of the **Alpine orogenic belt** along the southern coast of Spain (see Figure 11.5).

blue schist [110]: low-grade metamorphic rock formed under low-temperature, high-pressure conditions.

boudin, boudinage [57, 65]: a sausage-like shape produced as the result of the stretching of a layer; the process resulting in boudins (see Figure 4.8).

Briançonnais zone [93-94]: tectonic zone in the French-Swiss Alps thought to represent the small continental **Pennine terrane**, formerly situated on **Tethys Ocean** crust (see Figures 11.6, 7).

brittle (behaviour) [28-29]: failure (fracture) after no, or very little, **plastic** or **viscous** deformation when a **stress** is applied.

brittle–ductile transition [29]: the zone within the Earth's **crust** in which rock deformation changes from **brittle** above to **ductile** below; the depth at which this change takes place varies with different **lithologies** (see Figure 4.11).

Buchan (metamorphic zones) [101]: a sequence of metamorphic assemblages characterised by moderate pressure and steep temperature increases.

buckling [48]: a process of **folding** produced by compression acting approximately parallel to the folded layers (see Figure 6.6B).

C

Caledonian (orogeny), Caledonides [97]: a series of orogenic episodes affecting most of the British Isles together with western Norway and eastern Greenland during **Lower Palaeozoic** time, culminating in collision between the continents of **Laurentia**, **Baltica** and **Avalonia**; the belt affected by this orogeny (see Figures 12.1, 12.5).

Cape Smith belt [109]: fold/thrust belt within the **Eastern Churchill Province** of the **Trans-Hudson orogen** consisting of material thrust SE-wards onto the northern side of the **Superior craton**.

carbonates [81]: general term for sedimentary strata dominated by carbonate minerals, such as limestone and dolomite (see Table A3).

cataclasis, cataclastic [38]: deformation dominated by fracturing.

cataclastic flow [31]: mechanism of deformation of a crystalline solid, characterised by microscopic fracturing.

cauldron subsidence [70]: type of **permitted emplacement** of a **pluton**, marked by a central depressed block surrounded by a ring-shaped intrusion; the space for the intrusion has been created by the depression of the cylindrical block, which has sunk into the magma beneath (see Figure 8.4B).

Central Highland complex [101]: gneissose basement, correlated with the **Moine Supergroup**, underlying the **Dalradian Supergroup** in the north-east of the **Grampian Highlands** zone of the Scottish **Caledonides**.

central igneous complex [71]: group of associated igneous intrusions, including **stocks**, **ring dykes**, **cone sheets** and **radial dykes**, emplaced from a **magma chamber** at the base of a large volcano (e.g. see Figure 8.6).

channel flow [84]: gravity-induced lateral flow of a warm **ductile** layer in the lower **crust** of a thickened **orogen**, which is sandwiched between two stronger layers and squeezed towards the sides of the orogenic belt (see Figure 10.4).

chert [82]: a rock consisting of microcrystalline **silica**, in the form of sedimentary layers, or as nodules within limestones; formed either by the accumulation of microfossils or chemically, by precipitation.

chevron fold [50]: having straight **fold limbs** and a sharp angular **hinge** (see Figure 6.4E).

clastic (sedimentary rock) [116]: resulting from the disintegration of older rocks (see Table A3).

cleavage [57]: a general term covering a range of types of **foliation** in low-grade or un-metamorphosed rocks; e.g. **slaty**, **fracture**, and **crenulation cleavage**.

chlorite [60]: a family of complex hydrous magnesium-iron-aluminium silicates with a platy habit, found typically in low-grade metamorphic rocks.

co-axial strain [24]: progressive strain in which the orientations of the **principal strain axes** remain unchanged throughout the deformation; also known as **pure shear** (see Figure 4.6).

columnar jointing [42]: type of cooling **joint** found in lava flows or **sills** where the rock is divided into columns by sets of polygonal joints (see Figure 5.12B).

competent (layer, rock) [58]: relatively strong during deformation.

compressional stress [26]: positive directional **stress** tending to result in contraction.

compressive strength [39]: the threshold **stress** required for a body to fail under compression (see Figure 5.10).

concentric fold [48-49]: a fold in which the successive curved surfaces have a common centre of curvature (see Figure 6.5A).

cone sheet [71-73]: one of a concentric set of inclined dykes, with an arc-like outcrop, that lie on a set of conical surfaces converging approximately at a point source beneath a **central igneous complex** (e.g. see Figures 8.6, 7).

conjugate [98]: (faults, joints, shear zones) sets of planar structures with similar trends but opposed inclinations.

conservative (plate boundary) [14]: characterised by sideways motion against the adjoining plate along a **transform fault** (see Figures 3.7, 3.8).

constructive (plate boundary) [14]: where new plate is created by divergent motion of two adjacent plates, marked by an **ocean ridge** or **continental rift** (see Figures 3.7, 3.8).

continent [7]: in a geological context, the landmass of a continent plus adjacent sea bed underlain by **continental crust**, including the **continental shelf** and **continental slope** (see Figure 2.5).

continental crust [7]: that part of the **crust** underlying a continent, with variable composition corresponding, on average, to a mixture of granite and basalt.

continental drift [10]: the theory of the relative movements of continents around the Earth's surface.

continental platform [80]: the stable part of a continent, external to any **orogenic** activity (see Figure 10.1).

continental rift [15]: a **constructive plate boundary** situated on **continental crust** and marked by an elongate valley, usually volcanically active (see Figure 3.9B).

continental shelf [81]: that part of the continental margin covered by relatively shallow seas.

continental slope [81]: the zone between the **continental shelf** and the deep ocean.

convection current [19]: a pattern of flow in liquid or solid material driven by a temperature difference; this produces a density imbalance that provides the force necessary to generate the flow.

convergent plate boundary [1]: where opposing plates converge across the boundary – see **destructive plate boundary**.

cooling joints [42]: set of joints formed by contraction of an igneous body due to cooling (see Figure 5.12B).

Cordilleran (belt) [95-96]: **orogenic** belt situated in western North America resulting from a series of **subduction**-related events and collision of **terranes** during the **Mesozoic** and **Cenozoic Eras** (see Figures 2.2, 11.8).

core [5-6]: the innermost part of the Earth, from a depth of about 2900 km to the centre; the outer part is liquid (see Figure 2.3); also, the innermost part of a **fold**.

Core zone (terrane) [109]: tectonic zone within the **Early Proterozoic Eastern Churchill Province** consisting mainly of deformed and metamorphosed **Archaean** basement (see Figures 12.9, 12.10).

craton [9]: the stable part of a continental interior, unaffected by contemporary **orogenic** activity.

creep [28]: **visco-elastic** deformation resulting from a **stress** applied for a very long period of time.

crenulation cleavage [59]: type of **foliation** produced by sets of closely-spaced crenulations (microfolds), of the order of millimetres or less in width; when viewed at outcrop scale, it has the appearance of a set of bands that cut across the original layering (see Figures 7.1C, 7.2D, E).

cross joints [41]: set of joints oriented perpendicular to **fold axes**.

crust [60]: the uppermost layer of the Earth; see **oceanic crust** and **continental crust** (see Figure 2.4).

crystal lattice [30]: the regular arrangement of atoms making up the crystal framework.

Cumberland batholith [109]: large granitic **pluton** emplaced during the **Wathaman** phase of the **Hudsonian orogeny** in the **Meta Incognita craton,** SE Baffin Island (see Figure 12.9).

cuspate [65]: deformed wave-like surface between two materials of contrasting strength, consisting of rounded shapes separated by sharp 'cusps' which point towards the stronger material (see Figure 7.6D, 7.7B).

D

Dalradian Supergroup [99]: **Late Proterozoic** to **Cambrian** marine assemblage in the **Grampian Highlands** zone of the Scottish **Caledonides**.

debris apron [74]: a deposit of loose eroded rock debris, or **scree**, situated on the lower part of a steep slope, especially at the end of a gulley.

deep-focus (earthquake) [44]: originating at a depth of below 300 km.

deformation [20]: the process whereby rocks are physically altered by the effects of forces acting on them; such effects include **folds**, **faults** and **fabric**.

deformation twinning [31]: crystal twinning caused by deformation (see **twin gliding**) (see Figure 4.12C).

delta structure [61-63]: asymmetric shape of a deformed **porphyroblast**, named after the Greek letter delta (δ), used to determine the **shear sense** in a deformed **schist** or gneiss (see Figure 7.4E).

Dent Blanche nappe [94]: a **klippe** of crystalline basement of African parentage belonging to the **Apulian microplate;** it lies within the **Piémont zone** of the Western Alps but is considered to have rooted in the **Sesia Lanzo zone** (see Figure 11.7).

destructive plate boundary [16]: a **plate** boundary characterised by the convergent movement of adjoining plates, and by the destruction of oceanic plate or the collision of continents (see Figures 3.7, 3.8).

detachment horizon (or surface) [36, 49]: weak surface, usually parallel to bedding, used as a **thrust** plane or surface of discontinuity between two different folding modes.

dextral [25]: (of a **fault**, **shear zone** etc.) where the opposite side of a plane or zone moves to the right, as seen by an observer on one side.

Diablerets nappe [93]: part of the **Helvetic zone** of the French-Swiss Alps; a complex fold **nappe** underlain by a **ductile thrust**, directed north-westwards towards the European **foreland**. The sedimentary cover consists of **Mesozoic** platform sediments cored by crystalline European basement (see Figure 11.7).

diapir, diapirism [70]: deep-level **pluton** with a shape like an inverted tear drop (see Figure 8.1F); process of **forceful emplacement** of a diapir.

diffusion creep [31]: deformation mechanism in a crystalline material whereby **strain** is achieved in the solid state by diffusion through the crystal lattice or along grain boundaries from areas of high compressive **stress** to areas of low stress (Figure 4.12E).

dilation [22-23]: change in volume of a body; either positive (expansion) or negative (contraction) (see Figure 4.4A).

diorite [116]: coarse-grained intermediate igneous rock (see Table A2).

dip [32]: the angle, or direction, of inclination of a plane measured from the horizontal.

dip-slip fault [32]: a **fault** where movement has taken place up or down the fault surface (see Figure 5.2).

differential stress [29]: (or stress difference): the difference between the maximum and minimum stresses in a **stress field**.

dislocation creep [31]: mechanism of deformation achieved under high **pressure** and temperature by modifying the crystal shape by internal dislocations of the crystal lattice; features such as **deformation twins** and **kink bands** characterise crystals deformed in this way (see Figure 4.12B, C).

distortion [23]: type of **strain** involving shape change (see Figure 4.4B).

divergent plate boundary [1]: where opposing plates diverge across the boundary – see **constructive plate boundary**.

docking [85]: collision between a **terrane** or microplate with a continental **plate** (see Figure 10.5).

dome [53]: a structure whose surface dips radially, e.g. an **anticlinal pericline**.

drape fold [54]: **fold** situated above a **normal fault** and caused by the fault movement (see Figure 6.10B).

ductile [28]: (materials) that fail (fracture) only after experiencing considerable **plastic** or **viscous** deformation when a **stress** is applied.

dyke [66]: a body of igneous rock, with a sheet-like form and typically steep attitude, that cuts across the structure of the **host rock**, formed by magma filling a fissure (see Figure 8.1A).

dyke swarm [72]: a set of dykes originating from a magma source, e.g. a **central igneous complex**; the dykes may be arranged radially close to the source but become parallel at a distance from it (see Figure 8.7C).

dynamic [1]: (process) governed by forces or **stresses.**

E

earthquake [43]: a set of vibrations experienced at the Earth's surface, resulting from **fault** displacement or volcanic activity at depth.

earthquake intensity [43]: the severity of an earthquake as experienced at the surface, measured usually on the **Mercalli scale** (see Table 5.1).

earthquake magnitude [43]: the severity of an earthquake, measured as the amount of energy released, as calculated from the size of the resulting **earthquake waves**, using the **Richter scale** (see Table 5.1).

earthquake wave [43]: the set of vibrations travelling through the Earth, released at the source of an **earthquake**; these are of three main types: **primary (P-) waves, secondary (S-) waves** and **surface waves** (see Figure 5.14).

Eastern Churchill Province [108-109]: the NE branch of the **Early Proterozoic Trans-Hudson orogenic belt,** resulting from collision between the **North Atlantic craton** and the **Superior craton** (see Figures 12.9, 12.10).

ecc structure [61]: shorthand for **extensional crenulation cleavage** (see Figure 7.4A).

eclogite facies [84]: metamorphic assemblage formed under exceptionally high-pressure conditions.

effective pressure [38]: in a rock at depth, the **lithostatic pressure** minus the **pore fluid pressure**.

elastic (strain) [27]: where the strained body quickly returns to its unstrained shape on removal of the applied **stress**; 'temporary' strain.

elongation lineation [64]: linear **fabric** caused by the alignment of elongate objects (e.g. fossils or **porphyroblasts**) in a deformed rock.

English River domain [112]: tectonic zone in the **Superior Province** consisting of a ~60 km-wide E–W linear belt containing high-grade, ~2700 Ma-old, meta-sedimentary, meta-volcanic and granitic **gneisses** (see Figure 12.11).

Eo-Archaean [112]: division of the **Archaean Eon** 4,000-3,500Ma.

Eon [115]: primary unit of geological time (e.g. the **Phanerozoic**), subdivided into **eras** (see Table A1).

epicentre [43]: the point on the Earth's surface directly above the origin (**focus**) of an **earthquake**.

Era [115]: a unit of geological time (e.g. the **Palaeozoic**), subdivided into **Periods** (see Table A1).

eu-geosyncline [83]: term formerly used to describe a type of depositional assemblage within an **orogenic belt** containing thick sequences of predominantly **clastic** deposits together with volcanics and, importantly, **ophiolites;** now regarded as the product of a combination of deep ocean, **ocean trench** and **continental slope** environments.

G

evaporite [76]: a sedimentary deposit formed by the evaporation of warm shallow seas containing salts (especially sodium chloride) in solution.

extension (extensional stress) [22]: component of a **stress field** tending to produce elongation in a body.

F

fabric [26, 57]: a set of new structures, or a texture, produced in a rock as a result of deformation – e.g. **foliation** or **lineation**.

Farallon plate [13, 95]: oceanic **plate** east of the East Pacific ridge, now mostly subducted beneath the American plate (see Figure 3.6B).

fault [32]: a rock fracture across which appreciable movement has taken place.

fault breccia [38]: a fault rock consisting mainly (>30%) of coarse angular rock fragments.

fault gouge [38]: a fault rock with <30% visible fragmental material in a clay or silt matrix.

fault-plane solution [45]: method of analysing the pattern of first arrivals of **earthquake waves** to discover the orientation of the fault plane responsible for the earthquake.

finite strain [24]: the strain state at the end of a progressive deformation.

flat [37]: (of a **thrust**): that portion which follows bedding, as distinct from a **ramp**, which cuts through bedding (see Figure 5.7A).

flexural shear [50]: process of **folding** where the folded layers experience internal **shear strain** (see Figure 6.6C).

flexural slip [50]: process of **folding** where the folded layers slide past each other with little internal **strain** (see Figure 6.6A).

Flin Flon–Glennie (belt) [108]: oceanic volcanic-arc **terrane**, part of the **Western Churchill Province** (see Figures 12.9, 12.10).

focus [44]: (of an **earthquake**) the location of the origin.

fold [46]: a geological structure formed by the bending of a layer as a result of deformation.

fold amplitude [46]: half the 'height' of a fold, assuming the folded layer to be originally horizontal (see Figure 6.3D).

fold angle [46]: the angle enclosed by the tangents to the **fold limbs** (see Figure 6.3B).

fold axial plane [47]: the plane that bisects the **fold angle** (see Figure 6.3B).

fold axial surface [48]: the surface containing the **fold axes** in successive layers of a fold.

fold axis [47]: line of intersection of the fold surface and its **axial plane** (see Figure 6.3C).

fold closure [46]: see **fold hinge**.

fold hinge [46]: the zone of maximum change in orientation of a fold (also known as the 'fold closure') (see Figure 6.3A).

fold limb [46]: the part of a fold between adjacent **fold hinges** (see Figure 6.3A).

fold plunge [48]: the angle of inclination of the **fold axis**, measured from the horizontal, and its orientation.

fold wavelength [460]: in a folded layer, the distance between two adjacent **synforms** or **antiforms** (see Figure 6.3D).

Folded Jura [94]: tectonic zone consisting of folded and **thrust Mesozoic** sediments on the margin of the European **platform**, representing the outermost part of the **foreland fold-thrust belt** of the French-Swiss Alps; it is separated from the main part of the fold-thrust belt by the **foredeep** basin (see Figure 11.7).

foliation [57]: a set of planar structures produced in rock as a result of deformation.

footwall [22]: the lower side of a dip-slip fault.

force [21]: strictly, that which causes an acceleration in an object at rest or in uniform motion; the magnitude of the force equals the mass of the object times its acceleration; the cause of movement or deformation in a body.

forceful emplacement [68]: method of intrusion of an igneous body that relies on active deformation of the surrounding **host rock** to create space.

foredeep (basin) [80]: a sedimentary basin resulting from the depression of the **continental crust** due to the load of the rising **orogenic belt**; it contains a thick sequence of predominantly **clastic** sediments derived from the erosion of the main mountain range (Figure 10.1A).

foreland [80]: that part of the **continental crust** lying immediately adjacent to an **orogenic belt** and which has not been significantly affected by it (Figure 10.1A).

foreland basin [80]: see foredeep.

Foxe belt [108]: the north-eastern part of the Western Churchill Province resulting from collision between the Rae and Meta Incognita cratons during the Hudsonian orogeny (see Figures 12.9, 12.10).

fracture cleavage [57]: cleavage formed by closely spaced fractures.

G

gneiss [60-61]: a coarse-grained metamorphic rock characterised by the segregation of light-coloured minerals such as quartz and feldspar into bands or lenses separated by dark minerals such as micas or hornblende.

gneissosity [60-61]: the **foliation** characteristic of a **gneiss**.

Gondwana [10]: a **supercontinent** that existed during **Palaeozoic** time, consisting of the continents of South America, Africa, India, Antarctica and Australia (see Figure 3.1).

graben [34]: down-faulted block bounded by **normal faults** (see Figure 5.4A).

grain-boundary sliding [30]: mechanism of deformation involving movement along grain boundaries.

Grampian Highlands [99]: tectonic zone in the Scottish **Caledonides** dominated by **Late Proterozoic** to **Lower Palaeozoic** metasediments (see Figure 12.2).

Grampian orogeny [99]: early **Ordovician** tectono-thermal event, part of the **Caledonian orogeny**, responsible for the deformation and metamorphism of the **Grampian Highlands** of Scotland.

granite–greenstone terrain [111]: a type of early Precambrian terrain characterised by a predominance of granite interspersed with **greenstone belts**.

granular flow [30]: mechanism of deformation where grains flow freely past each other, as in unconsolidated sand.

granulite facies [84]: a metamorphic assemblage formed under high pressure and temperature.

gravity gliding [75]: the sliding of a sheet of rock (e.g. a **nappe**) under gravity (see Figure 9.1).

gravity spreading [75]: lateral spreading of the over-thickened **crust** of an **orogen** under gravity, leading to a reduction of crustal thickness; also known as '**orogen collapse**' (see Figure 9.3).

Great Glen fault [101]: major fault with a large **sinistral** component forming the boundary between the **Northern Highlands** and **Grampian Highlands** zones of the Scottish **Caledonides** (see Figure 12.2).

Greater Himalayan crystalline complex [89]: tectonic zone within the **Himalayan orogenic belt** consisting of high-grade **schists** and **gneisses** derived from **Late-Proterozoic** clastic sediments from the Indian **plate** (see Figure 11.3).

greenschist facies [101]: a metamorphic assemblage formed under low temperature and pressure.

greenstone belt [111]: a linear to irregularly-shaped assemblage of volcanic and sedimentary rocks situated within granitic terrain of early Precambrian age.

Grenville (belt, orogeny) [105]: a **Mid-Proterozoic orogenic belt** situated along the SE margin of North America; the orogeny responsible (see Figures 12.6, 12.7).

Grenville Front thrust [106]: major **ductile thrust** zone forming the NW boundary of the **Grenville orogenic belt** (see Figure 12.7).

greywacke [82, 116]: an unsorted sandstone composed of a variety of minerals; typically formed by deposition from a **turbidity current** (see Table A3).

Gulf of Aden rift [15]: an arm of the Indian Ocean caused by the separation of Arabia and Africa and linked with the Red Sea rift; part of a **constructive plate boundary** (see Figure 3.9B).

H

half-graben [99]: tilted fault block bounded by a single **normal fault** (see Figure 5.4C).

halite [76]: rock salt (sodium chloride).

hangingwall [32]: the upper side of a **dip-slip fault**.

Hearne craton [108]: Archaean block incorporated within the **Trans-Hudson orogen** (see Figure 12.9).

heave [32]: the horizontal separation between two formerly contiguous points achieved by a **dip-slip fault** (see Figure 5.3A, B).

Helvetic zone [94]: tectonic zone of the French-Swiss sector of the Alps consisting of a set of complex **fold nappes** underlain by **ductile thrusts**, directed towards the European **foreland**: they contain **Mesozoic continental platform** sediments cored by crystalline European basement (see Figure 11.7).

high-grade gneiss (terrain) [111]: type of early Precambrian terrain typified by highly deformed and metamorphosed, mostly gneissose, igneous and sedimentary rocks.

Highland Boundary fault [103]: major fault with a significant **sinistral** component forming the boundary between the **Grampian Highlands** and **Midland Valley** zones of the Scottish **Caledonides** (see Figure 12.2).

Himalayan Frontal thrust [88]: the edge of the **fold-thrust** belt on the Indian **foreland**, marking the southern margin of the central sector of the **Himalayan orogenic belt** (Figure 11.3).

Himalayan orogenic belt [87]: **orogenic belt** resulting from the collision between the Indian **plate** and Asia, and extending around the northern perimeter of the Indian sub-continent, including the Himalayas and neighbouring mountain ranges (see Figure 11.2).

horst [34]: uplifted block bounded by **normal faults** (see Figure 5.4A).

host rock [1]: the rock into which magma is intruded to form an igneous body.

hot spot [18]: a part of the Earth's **crust** exhibiting unusually high heat flow and vulcanicity, either within a plate (as in Hawaii) or on a **constructive (plate) boundary** (as in Iceland – see Figure 3.12).

Hudsonian [108]: an **Early Proterozoic orogeny** affecting large parts of North America including the **Trans-Hudson orogenic belt** (see Figure 12.9).

hydrostatic (pressure) [28]: state of stress in a fluid at rest, where the stress in all directions is equal.

I

Iapetus ocean [102]: **Lower Palaeozoic** ocean separating **Laurentia**, **Baltica** and **Avalonia** prior to collision during the **Caledonian** orogeny (see Figure 12.4).

Iapetus suture [104]: the boundary between the **Laurentian** and **Avalonian** plates, within the **Caledonian orogenic belt** (see Figure 12.2).

ice age [9]: term often used for the most recent (**Pleistocene**) glaciation, but can also refer to any glacial period in Earth history.

imbricate (structure, zone) [98]: where the same sequence of strata is repeated many times in successive **thrust** slices within a **nappe** (e.g. see Figure 5.7C–E).

incompetent (layer, rock) [58]: relatively weak during deformation.

Indus–Tsangpo suture [88]: boundary between the Indian and Asian **plates** (see Figure 11.2).

InSAR (Interferometric synthetic aperture radar) [41]: technique employing repeat measurements of ground features by aircraft- or satellite-mounted radar to detect small ground displacements over time periods of days to years.

Insubric line (fault) [94]: major **sinistral strike-slip fault** separating the **Ivrea** and **Sesia Lanzo zones** of the **Alpine orogenic belt** (see Figure 11.7).

interference structure [54]: structure caused by the superimposition of a later set of folds on an earlier set, producing complex shapes (see Figure 6.11).

intermediate-focus (earthquake) [44]: originating at a depth of between 60 and 300 km.

intermediate igneous rock [116]: an igneous rock intermediate in composition between acid and basic (see Table A2).

intersection lineation [64]: set of linear features (i.e. a **fabric**) caused by the intersection of two sets of planes (see Figure 7.6A).

intraplate [20, 44]: relating to the interior of a tectonic **plate** (rather than at a plate boundary); thus, e.g. 'intraplate earthquake'.

inversion [83]: process where a former extensional **normal fault** is re-used in a subsequent compressional event to become a **reverse fault** or **thrust** (see Figure 10.3).

island arc [5, 17]: a submerged arc-shaped mountain range, typically volcanic, situated alongside a **subduction zone** at a **destructive plate boundary** (see Figure 2.2).

isoclinal fold [46]: where both **fold limbs** are parallel or near parallel.

isostasy [7]: the theory or state of general gravitational equilibrium at the Earth's surface, in which topographic variations are balanced by density variations beneath (see Figure 2.4).

Isua supracrustals [112]: the earliest known rocks in the **North Atlantic craton**, a group of **supracrustal** volcanic and sedimentary rocks of **Eo-archaean** age (~3900 Ma) (see Figure 12.12).

Ivrea zone [94]: tectonic zone in the Italian sector of the **Alpine orogenic belt** consisting of crystalline basement of African parentage belonging to the **Apulian microplate** (see Figure 11.7).

J–K

joint [41]: a rock fracture across which there has been no significant movement.

Kaapvaal craton [110]: **Archaean craton** in South Africa stabilised by 2900 Ma.

Karakorum fault [41]: major dextral strike-slip fault in western Tibet thought to accommodate part of the northerly

movement of the Indian plate relative to Asia.

kinematic [1, 56]: (process) governed by movement.

kink band, kinking [31, 50]: a zone between two usually planar margins occupied by a set of straight **fold limbs**, which have rotated in such a way that they have slid laterally past each other with little internal **strain** (Figure 6.6F); the process of forming such a structure; a similar structure within a crystal.

klippe [94]: outcrop of a **thrust** sheet which has been separated from the rest of the sheet by erosion.

Knockan Crag [99]: locality in Assynt, NW Scotland, within the NW Highlands Geopark containing a visitor centre and geological trail with a section across the **Moine thrust**.

Knoydartian (orogenic phase) [99]: **Late Proterozoic** tectono-thermal event affecting the **Moine Supergroup** in the **Northern Highlands zone** of the Scottish **Caledonides**.

komatiite [110]: ultrabasic volcanic rock found especially in **Archaean greenstone belts** and held to indicate higher Archaean heat flow (see Table A2).

L

laccolith [68]: a **pluton** with an approximately lens-shaped form (see Figure 8.1C).

La Ronge–Lynn Lake (belt) [108]: oceanic volcanic arc **terrane**, part of the **Western Churchill Province** (see Figures 12.9, 12.10).

Lake District [104]: tectonic zone in the English **Caledonides** consisting of **Lower Palaeozoic** volcanics and **slate**-grade sediments, and interpreted as a volcanic arc situated at the northern margin of **Avalonia** (see Figure 12.2).

Laurasia [10]: a **supercontinent** that existed during **Upper Palaeozoic** time, consisting of the greater parts of the continents of North America, Europe and Asia (see Figure 11.1).

Laurentia [97]: a continent consisting of most of North America, Greenland and NW Scotland that existed during the

Lower Palaeozoic prior to the **Caledonian orogeny** (see Figures 12.1, 12.6).

Laurentian shield [95]: the stable Precambrian core of North America.

Leinster massif [104]: Lower Palaeozoic inlier in SE Ireland, including the Leinster granite, and representing the continuation of the **Lake District zone** of the English **Caledonides** (see Figure 12.2).

Lesser Himalayan schist belt [88]: tectonic zone within the **Himalayan orogenic belt** consisting of **slates** and **schists** derived from **Mid-Proterozoic clastic** sediments from the Indian **passive continental margin** and deformed into south-directed **fold-thrust** packages (see Figure 11.3).

Lewisian (complex) [98]: Archaean to **Late Proterozoic** basement (part of **Laurentia**), mostly composed of **gneisses**, of the NW **foreland** of the Scottish **Caledonides** (see Figure 12.2).

lineation [63]: a set of linear structures (a linear **fabric**) produced in rock as a result of deformation (see Figure 7.6).

listric fault [35]: a curved **normal fault** whose inclination decreases downwards (e.g. see Figure 5.4C).

lithology [27, 46]: the physical and chemical characteristics of a rock.

lithosphere [8, 15]: the strong upper layer of the Earth, with an average thickness of about 100 km, including the **crust** and part of the **mantle**; it consists of a number of **plates** that move over the weaker **asthenosphere** beneath (see Figure 2.4).

lithostatic pressure [22]: (sometimes termed 'confining pressure') the **stress field** exerted on a rock at depth caused by the weight of overlying rocks; since lithostatic pressure is exerted equally from all sides of a rock, it can result in an increase in density without a change in shape.

Longford Down massif [104]: continuation in Ireland of the **Southern Uplands** zone of the Scottish **Caledonides** (see Figure 12.2).

longitudinal joints [41]: set of joints oriented parallel to **fold axes** and perpendicular to the folded layer.

lopolith [68]: large (up to several hundred kilometres across), broadly concordant **pluton,** roughly lensoid or disc-shaped, and convex downwards (Figure 8.1D).

low-velocity layer [8]: region of the **mantle** at depths of ~100–200 km through which **seismic waves** travel more slowly than would be expected if it were composed of normal mantle material, indicating that its density is lower; it is therefore regarded as a zone of relative weakness and termed the **asthenosphere** (see Figure 2.4B).

M

Ma (mega-annus): time unit of a million years (see table A1).

magma chamber [16]: a large space within the **crust** into which magma is injected and subsequently may undergo magma differentiation; the igneous body resulting from the filling of a magma chamber is termed a **pluton** (see Figure 3.9A).

magnetic stripe [12]: a linear belt on the **oceanic crust** with the same magnetic polarity contrasting with adjacent stripes with the opposite polarity (see Figure 3.5).

Main Boundary thrust [88]: **fault** marking the southern boundary of the **Lesser Himalayan schist belt** (see Figure 11.3).

Main Central thrust [88]: major **ductile shear zone** forming the southern margin of the **Greater Himalayan crystalline complex** (see Figure 11.3).

Main Frontal thrust [88]: fault marking the outer margin of the fold-thrust belt of the central Himalayan orogenic belt.

Malene supracrustals [112]: group of **Meso-archaean supracrustal** rocks in the West Greenland sector of the **North Atlantic craton,** deposited on the **Amitsoq gneisses** and deformed with them ~3000 Ma ago.

mantle [5]: that part of the Earth's interior between the **crust** and the **core,** composed mainly of rock with an ultrabasic composition (see Figure 2.3).

mantle drag [20]: the force exerted on the base of a **tectonic plate** by the frictional drag of the underlying **mantle** (see Figure 3.14).

mantle plume [19]: a column of rising hot **mantle** material inferred to explain **hot spots** (see Figure 3.13B).

mantled gneiss dome [79]: dome-shaped area of granitic gneiss basement surrounded by an unconformable sedimentary cover.

Mercalli scale [43]: a subjective method of quantifying the intensity of an **earthquake** as experienced at a particular location at the surface on a scale of increasing severity I–XII (see Table 5.1).

Meso-Archaean [112]: division of the **Archaean Eon** 3500–3000 Ma.

Meta Incognita craton [109]: Archaean block incorporated within the **Trans-Hudson orogen,** dominated by the **Cumberland batholith** (see Figure 12.9).

Midland Valley [103]: tectonic zone in the Scottish **Caledonides** containing **Lower Palaeozoic ophiolites** and un-metamorphosed, gently folded **clastic** sediments; the zone is believed to represent an oceanic volcanic-arc **terrane** that has been tectonically welded to **Laurentia** (see Figure 12.2).

Midlands Platform [98]: the southern **foreland** of the British **Caledonides** consisting of the late Precambrian basement of **Avalonia,** mostly concealed by younger cover (see Figure 12.2).

mineral lineation [64]: linear **fabric** caused by the alignment of the long axes of a mineral (e.g. hornblende).

Minnesota River domain [112]: tectonic zone in the SW part of the **Superior Province,** within the USA, consisting of ~2700 Ma meta- volcanic and meta-sedimentary rocks on a ~3600 Ma **gneiss** basement (see Figure 12.11).

mio-geosyncline [83]: term formerly used for a type of depositional assemblage within an **orogenic belt** containing a thick sequence of predominantly **clastic** deposits of equivalent age to the much thinner **continental shelf** sequence; now regarded as the product of the **continental slope** environment.

mobile belt [9]: a zone of enhanced tectonic activity, e.g. an **orogenic belt**.

Moho(rovicic) (discontinuity) [6]: seismically defined base of the crust (see Figure 2.4).

Moine Supergroup [99]: unit of **Late Proterozoic** marine **clastic** meta-sediments in the **Northern Highlands** zone of the Scottish **Caledonides.**

Moine thrust (zone) [98-9]: major thrust (zone of thrusts) in NW Scotland defining the western boundary of the Caledonian orogenic belt (see Figure 12.3A).

molasse [94]: continental **clastic** deposits (often red) derived from erosion of an active mountain range.

Monte Rosa nappe [94]: part of the **Pennine zone** of the French-Swiss Alps: a complex **fold nappe** consisting of a core of crystalline basement, with an overthrust cover of **ophiolites** and ocean-floor sediments (see Figure 11.7).

Morcles nappe [94]: part of the **Helvetic zone** of the French-Swiss Alps; a complex **fold nappe** underlain by a **ductile thrust,** directed north-westwards towards the European **foreland**. The sedimentary cover consists of **Mesozoic continental platform** sediments cored by crystalline European basement; (see Figure 11.7).

mullion (structure) [64]: set of roughly semi-cylindrical structures resembling the architectural features of that name; they are typically formed at the interface between two folded layers of contrasting strength (see Figure 7.6E).

mylonite [38]: a fine-grained **fault** rock formed at depth under metamorphic conditions and typically showing a regular fine banding; associated with major **thrust** zones.

N

nappe [75]: a large sheet of rock, many kilometres in extent, resting on a basal fault, typically a **thrust**.

Neo-Archaean [112]: division of the **Archaean Eon** 3000–2500 Ma.

Neo-Tethys ocean [92]: oceanic region (now part of the Mediterranean Sea) formed during the late **Mesozoic** by the separation of **Apulia** from the African **plate** (see Figure 11.6).

New Quebec zone [109]**:** tectonic zone within the **Eastern Churchill Province** consisting mainly of strongly deformed and metamorphosed **Early Proterozoic supracrustal** rocks (see Figure 12.9).

normal fault [32]: a dip-slip fault whose upper side has moved down the fault plane (see Figure 5.2A).

normal stress [22]**:** the component of a **stress field** acting at right angles to a surface**.**

North Atlantic craton [108]**:** Archaean **craton** forming the NE **foreland** to the **Trans-Hudson orogen** (see Figure 12.9).

Northern Highlands [99]**:** tectonic zone in the Scottish **Caledonides** dominated by **Late Proterozoic** meta-sediments (see Figure 12.2).

Nuuk gneisses [113]**:** group of post-3000 Ma **granitic** intrusions, now gneisses, within the West Greenland sector of the **North Atlantic craton**, and affected by ~2800 Ma deformation and high-grade metamorphism.

O

oblate strain (ellipsoid) [23]**:** three-dimensional **strain** characterised by the minimum **strain axis** being much shorter than the maximum and intermediate strain axes, which are equal or near equal; 'flattening' strain (see Figure 4.5C).

oblique-slip fault [33]**:** a **fault** where movement has taken place obliquely on the fault plane.

ocean-floor spreading [11-13]**:** process of formation of new **oceanic crust** by the injection of magma at an **ocean ridge** (see Figures 3.5, 3.9A).

ocean ridge [5]**:** long, submerged oceanic mountain range, the site of a **constructive plate boundary** (see Figure 2.2).

ocean trench [5]**:** deep marine trough, site of a **subduction** zone and a **destructive plate boundary** (see Figure 2.2).

oceanic crust [7, 15]**:** crust composed mainly of basalt formed initially by volcanic activity at **constructive plate boundaries** such as **ocean ridges** (see Figure 3.9A).

oceanic plateau [85]**:** an especially thick and buoyant piece of **oceanic crust** that is less dense than normal and may become accreted to the upper **plate** of a **subduction zone**.

oil trap [35, 78]**:** a structure sealed by impermeable rock (e.g. salt or shale) that encloses an oil reservoir (see Figure 5.5B, C).

Old Red Sandstone [99]**:** unit dominated by non-marine **clastic** sediments derived from the erosion of the **Caledonian orogen** in Britain.

Oligocene (Epoch) [95]**:** a time period in the **Cenozoic**, 33.9–23.0 Ma ago.

oolith [25]**:** small calcite concretion in limestone formed in shallow warm seas.

open fold [46]**:** fold with a **fold angle** of more than 90°.

ophiolite [83]**:** a sequence of rock types interpreted as pieces of **oceanic crust**; a 'complete' sequence includes **ultra-basic** material assumed to be from the uppermost mantle, sheet-like basic and ultrabasic intrusions, a **sheeted dyke** layer, basalt **pillow lavas** and ocean-floor sediments.

orientation fabric [31]**:** where the crystallographic axes have a preferred orientation.

orogen collapse [76]**:** (also known as 'orogen spreading'): process of lateral spreading of an over-thickened **orogen** (see Figure 9.3).

orogenic belt, or orogen [18, 80]**:** part of the Earth's **crust**, typically a linear zone, that has undergone **orogenesis** (mountain building); orogenic belts are formed as a result of **plate** convergence, which causes crustal thickening, uplift and the forma-tion of mountains (see Figure 2.2).

orthogneiss [111]**: gneiss** of igneous origin.

Ottawan (orogenic phase) [107]**:** tectono-thermal phase dated at 1090–1020 Ma, responsible for the high-grade metamorphism of the interior of the **Grenville belt** and **thrust** movements on the **Allochthon Boundary thrust**.

overfold [48]**:** fold with an inclined **axial plane**, where one **fold limb** is inverted.

P

palaeomagnetism [11]**:** the study of the magnetic properties of rocks; principally to determine the orientation of their magnetic latitude and pole position at some point in the geological past.

Pangaea [10]**:** the **supercontinent**, consisting of the whole continental landmass, which existed during much of **Upper Palaeozoic** time (see Figure 3.2).

parallel fold [48]**:** a **fold** in which the folded layers are of approximately equal thickness throughout (see Figure 6.4C).

parasitic fold [53]**:** see **satellite fold**.

Parent–Spartan (belt) [108]**:** oceanic volcanic-arc **terrane**, part of the **Western Churchill Province** (see Figures 12.9, 12.10).

partitioning [86]**:** division of a **transtensional** or **transpressional stress** into perpendicular and **strike**-parallel components (see Figure 10.6).

passive bending [54]**: folding** caused by **faulting** and induced by the upward or downward movement of a layer under extension (e.g. a **drape fold**) (see Figure 6.10B).

passive continental margin [36]**:** a former **constructive plate boundary**, now inactive.

pelite, pelitic [117]**:** metamorphic rock formed from mudstone or shale (see Table A4).

penetrative lineation, foliation [64]**: fabric** that penetrates throughout the rock.

Pennine terrane [92]**:** a small continen-tal **terrane**, formerly situated on **Tethys Ocean crust**, consisting of a **Mesozoic** sedimentary sequence distinct from that of the **foreland**, and underlain by crystal-line basement (see Figure 11.6).

Penrhyn–Piling basin [108]**: early Pro-terozoic** (see Table A1) sedimentary basin situated at the margin of the **Rae craton** within the **Trans-Hudson orogen** (see Figure 12.10A).

pericline [53]**: fold** which varies in height along its length such that the **fold plunge** is in opposite directions at each end, and may be either **anticlinal** or **synclinal** (see Figure 6.1).

peridotite [116]**:** an ultrabasic rock containing a high proportion of pyroxene and olivine (see Table A2).

Period [115]**:** a unit of geological time (see Table A1).

permitted emplacement [68]**:** process of intrusion of an igneous body that is aided by the **host rock** moving aside to create space.

Phanerozoic (Eon) [115]**:** the unit of geological time from the beginning of the **Cambrian Period** (542 Ma ago) to the present (see Table A1).

phenocryst [25]**:** large crystal in an igneous rock, set in a finer-grained matrix.

phyllite [60, 117]**:** type of fine-grained metamorphic rock typically containing muscovite or chlorite, derived from mudstone or shale, and intermediate in metamorphic grade between **slate** and **schist** (see Table A4).

Piémont zone [95]**:** tectonic zone in the Western Alps containing **ophiolites** and ocean-floor sediments representing pieces of **Tethys Ocean crust** and upper **mantle**, taken to indicate the presence of a **suture** between the **Pennine terrane** and **Apulia** (see Figure 11.7).

Pikwitonei domain [111]**:** tectonic zone at the NW margin of the **Superior Province;** a **high-grade gneiss** belt including 3400–3000 Ma gneisses affected by **granulite-facies** metamorphism and deformation at 2716–2642 Ma (see Figure 12.11).

Pilbara craton [110]**:** Archaean craton in NW Australia stabilised at 2800 Ma.

Piling Group [108]**:** **supracrustal** cover sequence on the **Rae craton, Trans-Hudson orogen**; the lower part consists of shallow-marine shelf sediments and is overlain by **foredeep** deposits.

pillow lava [83]**:** lava deposited in water, especially on the ocean floor, and characterised by structures resembling pillows and tubes.

plane strain [23]**:** three-dimensional **strain** characterised by the intermediate **principal strain axis** remaining unchanged (see Figure 4.5D).

plastic (strain) [27]**:** permanent **strain,** remaining after the deforming **stress** is removed.

plate [14]**:** a relatively stable piece of the **lithosphere** that moves independently of adjoining plates; plate boundaries are of three types, **constructive, destructive** and **conservative** (see Figures 3.7, 3.8).

plate tectonics [13]**:** the theory that ascribes tectonic processes to the relative movement of the **lithosphere plates**.

Plateau Jura [94]**:** part of the **foreland** of the French-Swiss sector of the **Alpine orogenic belt,** characterised by undeformed **Mesozoic** strata (see Figure 11.7).

Pleistocene (Epoch) [9]**:** subdivision of the **Quaternary Period**, characterised by multiple glaciations, 1.64–0.01 Ma ago.

plug [66]**:** an approximately cylindrical igneous intrusion filling a volcanic **vent** (see Figure 8.1B).

plunge [48]**:** the angle of inclination, measured from the horizontal, and/or the orientation, of a **fold axis** or **lineation**.

pluton [66, 70]**:** an igneous body of large dimensions, both horizontally and vertically.

Po basin [94]**:** tectonic zone in the Italian sector of the Western Alps **orogen**; it consists of non-marine **clastic** sediments derived from the rising Alps and deposited in a **foredeep basin** on the **Apulian plate** (see Figure 11.7).

pore-fluid pressure [28]**:** the pressure exerted by the pore fluid in a rock at depth, which partially or completely equals the **lithostatic pressure**, depending on the proportion of fluid in the rock.

porphyroblast [62]**:** a large crystal in a metamorphic rock within a finer-grained matrix.

Pre-Alps [94]**:** tectonic zone of the French-Swiss Alps **orogen** consisting of a **klippe** of **Piémont zone** rocks lying within the **foreland basin** and **Helvetic zones** (see Figure 11.7).

preferred orientation [31]**:** where a set of objects (e.g. crystals) in a rock have, or tend towards, a parallel orientation **fabric** caused by deformation.

pressure solution [31]**:** deformation process where material is dissolved in regions of higher **stress** and re-deposited in regions of lower stress (see Figure 4.12F).

pressure solution cleavage [57]**:** **foliation** produced by **pressure solution**.

primary (P-) waves [44]**:** the first set of **earthquake waves** to arrive at a recording station; they travel through the Earth by a process of alternate expansion and compression of the material through which they are transmitted in the same way as sound waves (see Figure 5.14A).

principal strain (axes) [23]**:** the mutually perpendicular greatest, least and intermediate **strain** directions in a deformed body (see Figure 4.5).

principal stress (axis) [23]**:** the mutually perpendicular greatest, least and intermediate **stress** directions in a deformed body (see Figure 4.2).

prolate strain (ellipsoid) [23]**:** three-dimensional **strain** characterised by the maximum **principal strain axis** being much larger than the intermediate and minimum strain axes, which are equal or nearly equal; elongation strain (see Figure 4.5B).

Proterozoic Eon [115]**:** unit of geological time between 2500 and 542 Ma; the upper part of the **Precambrian** (see Table A1).

Psammite [117]**:** metamorphosed sandstone (see Table A4).

pseudotachylite [31]**:** **fault** rock, formed by frictional melting, in the form of intrusive veins (see Figure 5.9B).

pure shear [24]**:** see **co-axial strain**.

Purtuniq ophiolite [109]**:** a 5 km-thick **ophiolite** sequence within the **Cape Smith belt** of the **Trans-Hudson orogen**.

Q-R

Quaternary (Period) [115]**:** unit of geological time in the **Cenozoic Era**, 2.5–0 Ma (see Table A1).

Quetico domain [112]**:** tectonic zone in the **Superior Province** consisting predominantly of ~2700 Ma-old meta-**greywackes** believed to be derived from the adjoining volcanic domains (see Figure 12.11).

radial dykes [72]**:** a set of **dykes** radiating

from a central point in a **central igneous complex** (see Figure 8.7).

radiolarian chert [83]**:** a **siliceous** deposit formed from the skeletons of radiolaria (unicellular marine organisms) and found in deep-ocean environments.

Rae craton [108]**:** Archaean **craton** forming part of the NW **foreland** of the **Trans-Hudson orogen** (see Figure 12.9).

raised beach [9]**:** an old beach above the present shore line formed at a time of higher sea level (see Figure 2.7).

ramp [36]**:** that part of a **thrust fault** which cuts up through bedding (see Figure 5.7A).

recumbent fold [48]**:** having a horizontal or gently inclined **axial plane** with one **fold limb** inverted.

Red Sea rift [16]**:** an elongate depression, occupied by the Red Sea, caused by the separation of Arabia from Africa, and floored by **oceanic crust**; part of a **constructive plate boundary** (see Figure 3.9B).

regional metamorphism [101]**:** metamorphism generated by regional changes in temperature and/or pressure, caused by **orogenic** processes.

Reindeer zone [108]**:** that part of the **Trans-Hudson orogen** between the **Hearne** and **Superior cratons**.

Reindeer–Foxe phase [111]**:** part of the **Hudsonian orogeny** (~1880–1865 Ma), marked by collisions between **Meta Incognita** and **Rae cratons**, formation of the **Torngat belt** and accretion of oceanic arc **terranes** to the **Hearne craton** (see Figure 12.10).

Renbu–Zedong thrust [89]**:** fault cutting the **Indus–Tsangpo suture** in the central sector of the **Himalayan orogenic belt**, and which forms the boundary between the Indian and Asian plates (Figure 11.3).

reverse fault [32]**:** a **dip-slip fault** whose upper side has moved up the fault surface (see Figure 5.2B).

Rhinns complex [101]**:** **Early Proterozoic** basement (part of the **Laurentian foreland**) in the **Grampian Highlands** zone of the Scottish **Caledonides**.

Richter scale [43]**:** scale for measuring

earthquake magnitude, numbered from 1 (lowest) to 10 (highest), where each unit represents an increase of x10 in energy released (see Table 5.1).

ridge-push force [20]**:** the force exerted on a **tectonic plate** by the gravitational push exerted by an **ocean ridge** (see Figure 3.14).

Rigolet (orogenic phase) [107]**:** tectonothermal phase, dated at 1090–980 Ma, responsible for the **Grenville Front thrust** and related metamorphism and deformation in the outer zone of the **Grenville belt**.

ring dyke [72]**:** steeply inclined ring-shaped or arc-shaped intrusion surrounding a **central igneous complex** (see Figures 8.4, 8.6, 8.7).

rodding (lineation) [64]**:** linear **fabric** on a centimetre scale caused by the alignment of elongate features in a deformed rock.

Rodinia [105]**:** a **Late Proterozoic supercontinent** consisting of most of the then-existing continental masses, assembled during **Mid-Proterozoic orogenies** (see Figure 12.6).

rollover [34]**:** a **hangingwall antiform** (see Figure 5.4B).

rotational strain [24]**:** progressive strain in which the **principal strain axes** rotate during the deformation; also known as **simple shear** or 'non-co-axial' strain.

S

salt diapir [77]**:** **diapiric** salt intrusion (see Figure 9.4).

salt dome [77]**:** a dome-shaped structure formed by the upward migration of salt under gravitational pressure from its source layer (see Figure 9.4).

salt pillow [78]**:** an elongate **anticlinal** body of salt formed as above.

salt plug [77]**:** a roughly cylindrical salt intrusion.

salt wall [78]**:** a steep-sided sheet of salt, intruded (e.g.) up a **fault**.

San Andreas fault (zone) [37, 45]**:** a major **transform fault** zone situated along the west coast of California, separating the Americas **plate** from the Pacific plate (see Figures 3.6B, 3.7, 5.8A).

Sashigo domain [111]**:** tectonic zone in

the **Superior Province** containing thin, isolated **greenstone belts** surrounded by extensive areas of granite and **gneiss**; the greenstone belts overlie >3000 Ma-old granitic basement (see Figure 12.11).

satellite fold [53]**:** (or parasitic fold) a smaller fold situated on the **fold limb** of a larger fold (see Figure 6.9).

S-C structure [61]**:** type of asymmetric **fabric** formed by a set of narrow **shear zones** (C planes) cutting the **schistosity** (S planes) obliquely (see Figure 7.4B, 7.5B).

Scandian (orogenic phase) [99]**:** major early **Silurian** tectono-thermal event affecting the **Northern Highlands** zone of the Scottish **Caledonides**; attributed to the collision between **Baltica** and **Laurentia**.

schist [61]**:** a medium- or coarse-grained metamorphic rock characterised by the parallel alignment of platy minerals such as chlorite or mica (see Figure 7.3A, B).

schistosity [59]**:** the **foliation** characteristic of a **schist**.

scree [75]**:** loose rock debris formed at the foot of a steep slope.

sea-floor spreading [10]**:** the theory of the generation of **ocean crust** at **ocean ridges** and its destruction at **ocean trenches** (see Figure 3.4).

secondary (S-) waves [44]**:** the second set of **earthquake waves** to arrive at a recording station; they are transmitted through the Earth by a process of lateral vibrations (i.e. **shear**) of the material through which they travel (see Figure 5.14A).

seismic, seismicity [8]**:** relating to **earthquake** activity.

seismic section [35]**:** cross-section through the crust obtained from artificial seismic sources to elucidate sub-surface structure, usually for prospecting purposes (see Figure 5.5A).

seismic wave [44]**:** see **earthquake wave**.

seismograph [44]**:** instrument for detecting **earthquake waves**.

semipelite [117]**:** metamorphic rock derived from siltstone (see Table A4).

Sesia Lanzo zone [94]**:** tectonic zone in the Italian sector of the **Alpine orogenic belt** consisting of crystalline basement of

African origin belonging to the **Apulian (micro-)plate**; northern continuation of the **Ivrea zone** (see Figure 11.7).

Sgurr Beag thrust [99]: major **ductile thrust** affecting the **Moine Supergroup** in the **Northern Highlands** zone of the Scottish **Caledonides**.

shallow-focus (earthquake) [44]: originating at a depth of 0 to 60 km below the surface.

shape fabric [31]: fabric caused by the alignment of the flattened or elongate shapes of grains or grain aggregates in a deformed rock.

shear [2]: (applied to a plane, e.g. a fracture or **fault**), process resulting from an oblique **stress** tending to produce opposed directions of movement on either side of the plane (see Figure 4.3).

shear failure envelope [39-40]: line representing the variation in conditions for shear failure on the **shear stress/ normal stress** diagram; enclosing the region of stable **stress** states for a given material (see Figure 5.10C).

shear sense [61]: the direction of rotation of an object affected by **shear strain** (e.g. **sinistral** or **dextral**) (see Figure 7.4).

shear strain [23]: rotational **strain** in which the strain axes progressively rotate during the deformation (see Figure 4.4C).

shear strength [39]: the threshold **stress** required for failure in a body subject to **shear stress** (see Figure 5.10).

shear stress [22]: the component of a **stress field** acting parallel to a surface of a body and tending to produce a rotation of the body (see Figure 4.3B).

shear zone [54]: a zone of **ductile** deformation between two rock masses moving in opposite directions; the equivalent of a **fault** at depth (see Figure 6.13).

sheeted dyke (layer) [16]: that part of the **oceanic crust** composed entirely of steep **dykes** that have acted as feeders to the basalt layer above (see Figure 3.9A).

sigma structure [63]: asymmetric shape of a deformed **porphyroblast**, named after the Greek letter σ, used to determine the **shear sense** in a deformed **schist** or **gneiss** (see Figure 7.4D).

sill [66]: a sheet-like igneous body typically with a horizontal or gently inclined attitude formed by the injection of magma along a fissure parallel to the prevailing structure (e.g. bedding) of the **host rock** (see Figure 8.1A).

siltstone [116]: clastic sedimentary rock composed mainly of fragments between 0.004 and 0.5mm in diameter (see Table A3).

Silurian (Period) [115]: unit of geological time in the Palaeozoic Era 444-416 Ma ago (see Table A1).

simple shear [24]: see **rotational strain**.

similar fold [48]: where the folded layers vary in thickness but each layer has the same shape (see Figure 6.4D).

sinistral [24]: (of a **fault**, **shear zone** etc.): where the opposite side of the structure moves to the left as seen by an observer on one side.

Siwalik Formation [88]: sequence of un-metamorphosed sediments formed in the **foredeep basin** of the **Himalayan orogenic belt** and derived from the rising Himalayan mountains (see Figure 11.3).

slab-pull force [20]: the force exerted on a **tectonic plate** by the gravitational pull of a sinking **lithosphere** slab at a **subduction zone** (see Figure 3.14A).

slate [117]: a fine-grained metamorphic rock formed from shale or mudstone and characterised by **slaty cleavage (see Table A3).**

slaty cleavage [58]: the closely-spaced **foliation** characteristic of a **slate** (see Figure 7.1A, B).

Slave craton [108]: Archaean **craton** forming part of the NW **foreland** of the **Trans-Hudson orogen** (see Figure 12.9).

slickenfibre [33, 64]: elongate crystals formed by progressive movement along a **fault** plane and indicating the direction of movement (see Figure 7.6C).

slickenline [33, 64]: grooves formed on a **fault** surface by the grinding action of the opposite side of the fault and indicating the direction of movement (see Figure 7.6C).

slickenside (surface) [33]: polished surface formed by movement along a **fault** plane (see Figure 7.6C).

Snow Lake belt [110]: **Early Proterozoic** volcanic arc on the western margin of the **Superior craton** (see Figure 12.10A).

Snowbird orogenic phase [110]: early **orogenic** phase (~1920–1890 Ma) of the **Hudsonian orogeny,** marked by the collision of the **Hearne** and **Rae cratons**.

soil creep [75]: process caused by the flow of small rock or soil particles down a steep slope under gravity, causing (e.g.) bedding to rotate down-slope (Figure 9.1C).

solution creep [31]: deformation mechanism whereby material is transferred from areas of high **compressive stress** to areas of low stress by solution and re-deposition from a fluid (see also **pressure solution**) (Figure 4.12E, F).

Solway suture [104]: concealed fault between the **Southern Uplands** and **Lake District** zones, separating the Scottish and English **Caledonides**; the boundary between **Laurentia** and **Avalonia;** part of the **Iapetus suture** (see Figure 12.2).

South Tibetan detachment [89]: north-dipping **normal fault** marking the lower boundary of the **Tethyan shelf sequence** in the Central Himalayas (see Figure 11.3).

Southern Alps [92]: branch of the **Alpine orogenic belt** along the southern side of the Central and Eastern Alps, consisting of a **fold-thrust** belt directed southwards towards the **Apulian microplate** (see Figure 11.5).

Southern Uplands [103]: tectonic zone in the Scottish **Caledonides** dominated by a **Lower Palaeozoic** assemblage consisting mostly of **greywackes** and **slates**, and interpreted as an **accretionary prism** (see Figure 12.2).

Southern Uplands fault [103]: major fault separating the **Midland Valley** and **Southern Uplands** tectonic zones of the Scottish **Caledonides** (see Figure 12.2).

space problem [69]: the problem of determining how space is made to accommodate a large **pluton** within the **crust**.

spaced cleavage [59]: any type of **cleavage** in which the individual **fabric** planes are obviously separated by zones of un-cleaved rock; normally applied to cleavages seen at outcrop scale (e.g. see Figure 7.1D).

splay fault [42]: fracture formed at an angle to an existing **fault** at its termination (see Figure 5.13B).

St. Bernard nappe [94]: large complex **ductile nappe** within the **Briançonnais zone** of the French-Swiss Alps, consisting of **Mesozoic continental shelf** sediments and cored by European crystalline basement (see Figure 11.7).

staircase path [36]: **thrust** plane geometry marked by alternations of **ramps** and flat sections (see Figure 5.7A).

stock [70]: intermediate-sized **pluton**, with broadly cylindrical shape and discordant relationships to the **host rocks**, typically formed at high levels of the **crust** (see Figure 8.1E).

Stoer Group [98]: **Late Proterozoic** non-marine **clastic** sedimentary sequence on the northern **foreland** of the **Caledonides** in NW Scotland.

Storegga slide [75]: very large slope failure off the coast of Norway, dated at 6100 BCE, believed to have triggered a huge tsunami in the North Atlantic.

stoping [70]: process whereby an igneous **pluton** makes space for its upwards ascent by prising off blocks of its roof, which then sink into the magma (see Figure 8.4D).

strain [21]: change in shape produced by a **stress** or **stress field**.

strain ellipsoid [23]: a geometrical device to illustrate the three-dimensional variation of **strain** (as if resulting from deformation of a sphere) in a deformed body (see Figure 4.5).

strain gauge [31]: instrument for measuring strain within a rock body by recording small changes in dimension.

strain marker [25]: an object within a rock body whose shape can be used to help to determine the **strain** of the whole body.

stress [21]: a pair of equal and opposite **forces** acting on unit area of a surface (Figure 4.1A).

stress field [21]: system of **stresses** acting in three dimensions on a body.

strike [1]: (of an inclined plane, e.g. a **fault**, bed, etc.): the orientation of a horizontal line within the plane.

strike-slip (movement, fault etc.) [32]: where the movement has taken place horizontally along a fault or plane (see also wrench fault – see Figure 5.2C).

structure [1]: a body of rock whose shape can be defined geometrically and which has originated by a geological process, e.g. a **fold**, **fault** or igneous intrusion.

Sub-Briançonnais zone [94]: tectonic zone in the French-Swiss Alps lying structurally beneath the **Briançonnais zone** and dominated by marine **clastic** sediments thought to have formed in an **ocean trench** between the European **plate** margin and the **Pennine terrane** (see Figure 11.7).

subduction [16]: the process whereby an oceanic **plate** descends into the **mantle** along a **subduction zone**; part of the subducted plate melts to give rise to a zone of volcanoes on the opposite (upper) plate (see Figures 3.10, 3.11).

Sugluq craton [108]: Archaean **craton** incorporated within the **Trans-Hudson orogen** (see Figure 12.9).

supercontinent [10]: a large continental mass consisting of several components that previously, or subsequently, were themselves continents.

Superior Province (or craton) [111]: large Archaean **craton** occupying a major part of Canada east and south of Hudson Bay and extending into northern USA (see Figure 12.9).

supracrustal (rocks) [00]: deposited on the surface, i.e. sediments and volcanics.

surface waves [44]: set of **earthquake waves** that travel around the surface of the Earth's **crust** and arrive at a recording station after the **P-** and **S-waves** (see Figure 5.14A).

suture [83]: the boundary between two continental **plates** brought together during collisional **orogeny**.

Swordly nappe [99]: nappe in the eastern part of the **Northern Highlands zone** of the Scottish **Caledonides**, composed of metasediments of the **Moine Supergroup**.

symmetric fold [49]: where the **fold limbs** are of equal length (see Figure 6.4F).

syncline [48]: a **fold** where the younger beds are in the **core** (see Figure 6.4A).

synform [48]: a **fold** that closes downwards (see Figure 6.4A).

synthetic shears [62]: set of narrow shear zones making a shallow angle to the main **foliation** and with the same **shear sense** (see Figure 7.4B).

T

Taltson-Thelon belt [108]: an **Early Proterozoic orogenic belt** forming part of the **foreland** of the **Trans-Hudson orogen** (see Figure 12.9).

tectonic [1]: (of a process or structure) relating to, or caused by, major Earth forces e.g. **plate tectonics**.

tectonic regime [1]: a set of structures or processes caused by or related to the same overall tectonic system; e.g. **compressional**, **extensional** or **strike-slip** tectonic regimes.

tensile strength [39]: the threshold **stress** required for a body to fail under **extension** (see Figure 5.10).

terrane [85]: a piece of **crust**, smaller in scale than a **plate**, and now part of an **orogenic belt**, which exhibits a different structural history from neighbouring regions and may contain **palaeomagnetic** or fossil evidence of derivation from some distance away (see Figure 10.5).

Tethyan shelf sequence [88]: tectonic zone within the **Himalayan orogenic belt** consisting of largely un-metamorphosed **Cambrian** to **Palaeogene** marine strata, mainly carbonates and shales, originally laid down on the Indian **continental shelf** (see Figure 11.3).

Tethys ocean [92]: the ocean that separated **Laurasia** and **Gondwana** during **Upper Palaeozoic** and early **Mesozoic** time (see Figures 3.2, 11.1).

throw [32]: the vertical displacement of a **fault** (see Figure 5.3A).

thrust fault [36]: a **reverse fault** that is

generally gently inclined (<45°) although it may become steepened by subsequent movements (see Figure 5.7).

tight fold [46]: fold with a **fold angle** of less than 90°.

tonalite [116]: coarse-grained **acid** igneous rock characterised by an absence of alkali-feldspar (see Table A2).

Torngat belt [109]: tectonic zone within the **Eastern Churchill Province** consisting mainly of strongly deformed and metamorphosed **Early Proterozoic supracrustal** rocks (see Figure 12.9).

Torridonian (Supergroup) [98]: a **Late Proterozoic** non-marine sedimentary sequence on the NW **foreland** of the **Caledonides** in NW Scotland.

Trans-Hudson orogenic belt [107]: **Early Proterozoic** belt, centred on Hudson Bay, formed during the **Hudsonian orogeny**; it extends from Wyoming in the northern USA to Baffin Island in northern Canada (see Figure 12.9).

transform fault [14]: a fault that forms part of a **plate** boundary where the plates on each side move in opposite directions, parallel to the trend of the fault (see Figures 3.7, 3.8).

transpression [26]: combination of **compressional** and **shear stress**.

transtension [26]: combination of **extensional** and **shear stress**.

triple junction [15]: meeting point of three **constructive plate boundaries**, one of which may be inactive (a 'failed arm') (e.g. see Figure 3.9B).

tsunami [44]: water waves with long wavelengths caused by sudden movements of the sea floor, e.g. caused by **earthquakes**; they travel at high speed and may cause great destruction when they reach land.

turbidite [101]: a sedimentary deposit formed by a **turbidity current,** characterised by a range of **clastic** particle sizes and compositions.

turbidity current (or flow): a water current generated by gravity-induced flow, carrying large quantities of sediment of varying coarseness in suspension.

twin gliding [31]: crystal twinning caused by deformation (see Figure 4.12C)**.**

U

Uchi domain [112]: tectonic zone in the **Superior Province** consisting of a series of thin, tightly-folded, E–W-trending belts of meta-volcanic and meta-sedimentary rocks intruded by **granodioritic** to **tonalitic plutons** (see Figure 12.11, Table A2).

unroofing (joints) [42]: (formed by) **stress** release caused by removal of gravitational load by erosion (see Figure 5.12A).

upright fold [48]: fold with a steeply inclined **axial plane**, where neither **fold limb** is inverted.

V

vent [66]: the mouth of a volcano; or the body feeding a volcano (see Figure 8.1B).

Virtual Seismic Atlas (VSA): repository of **seismic sections** hosted by the University of Leeds to illustrate various types of geological structure.

viscosity [28]: the quality of stickiness or fluidity of a material, measured by its rate of flow.

visco-elastic (strain) [28]: permanent **strain** with an initial **elastic** component (see Figure 4.10).

W

Wathaman phase [110]: of the **Hudsonian orogeny** (~1865–1840 Ma) marked (e.g.) by accretion of volcanic arc **terranes** to the **Sugluq** and **Hearne cratons** and emplacement of granite **batholiths** (see Figure 12.10).

Wawa domain [112]: tectonic zone in the **Superior Province** of Canada; the western extension of the **Abitibi domain;** a **granite–greenstone terrain** consisting of a late Archaean **subduction–accretion** complex of intra-oceanic and volcanic island-arc belts (see Figure 12.11).

Welsh Basin [104]: tectonic zone in the British **Caledonides** dominated by **Lower Palaeozoic** volcanics, **greywackes** and **slates**, interpreted as a **back-arc basin** situated on the **Avalonian** plate (see Figures 12.2, 12.4).

Western Churchill Province [108]: the western branch of the **Trans-Hudson orogenic belt**, extending in a NE direction from Wyoming to Baffin Island (see Figure 12.9).

Wildhorn nappe [94]: part of the **Helvetic zone** of the French-Swiss Alps; a complex **fold nappe** underlain by a **ductile thrust**, directed north-westwards towards the European **foreland**. The sedimentary cover consists of **Mesozoic continental shelf** sediments underlain by crystalline European basement (see Figure 11.7).

Winnipeg River domain [112]: tectonic zone in the Archaean **Superior Province** of Canada consisting of a **granite–greenstone terrain** dominated by **granitic** rocks; it includes 3319 Ma **tonalitic gneisses**, highly metamorphosed **supracrustal** rocks and ~2700 Ma **granodioritic** to **dioritic plutons** (see Figure 12.11, Table A2).

Wrangellia terrane [95]: displaced **terrane**, part of the **Cordilleran orogenic belt** of North America, which accreted to the **Alexander terrane** during the **Jurassic**; the combined terranes collided with **Laurentia** in the **Cretaceous** (see Figure 11.8).

wrench fault [37]: a **fault** where the displacement is horizontal and parallel to the trend of the fault; a **strike-slip** fault (see Figure 5.2C).

Y

yield strength [28]: the threshold **stress** required for a material to deform by **plastic** or **viscous strain**.

Further reading

More advanced textbooks recommended for further study
Analysis of Geological Structures (1990) by N.J. Price & J.W. Cosgrove, Cambridge University Press.
Foundations of Structural Geology, 3rd edition (1997), by R.G. Park, Chapman & Hall (now Routledge).
Structural Geology, 2nd edition, (2007) by R.J. Twiss & E.M. Moores, Freeman.
Structural Geology (2010) by Haakon Fossen, Cambridge University Press.
Tectonics (1995) by E.M. Moores & R.J. Twiss, Freeman.

Web-based resources
Learn Structure and *Assynt Geology* by Robert Butler: excellent on-line resources for students.

Also useful:
Virtual Seismic Atlas: Robert Butler, University of Leeds Earth Science Dept.

Finally: put *structural geology* into a search engine and see what comes up!